IMAGE CAPTIONS, DUST JACKET

1. This sequence of still images from the GOES-17 satellite shows the plume of the Hunga Tonga—Hunga Ha'apai submarine volcano eruption at various stages on January 15, 2022. NASA Earth Observatory images by Joshua Stevens, using data courtesy of Kristopher Bedka and Konstantin Khlopenkov/NASA Langley Research Center. GOES-17 imagery courtesy of NOAA and the National Environmental Satellite, Data, and Information Service (NESDIS).

2. Operation Ivy nuclear weapons test, 1 November 1952, from *Trinity and Beyond: The Atomic Bomb Movie*, dir. Peter Kuran (1995; Thousand Oaks, California: Visual Concept Entertainment, Documentary Film Works).

3. Perpendicular changes to land caused by the Great Kantō earthquake, Sagami Trough, Japan, 1923, from *The Reconstruction of Tokyo* (Tokyo: Tokyo City, 1933).

4. Satellite image of Bikini Atoll, Marshall Islands, 2023. NASA Earth Observatory image by Jesse Allan and Robert Simmon using Landset data from the U.S. Geographical Survey.

5. LIGO Hanford Observatory, Richland, Washington.

6. Japanese woodblock print depicting people attempting to capture Namazu, the giant catfish that was believed to cause earthquakes, c. 1855/56. University of Columbia Library—Rare Books and Special Collections.

IMAGE CAPTIONS, ENDSHEETS

Front: Tarvurvur volcano eruption, Papua New Guinea, 19 February 2010. Photograph by James Chisholm.

Back: See no. **2**.

This catalogue was published in conjunction with the exhibition *Energy Fields: Vibrations of the Pacific*, curated by Lawrence English and Robert Takahashi Novak and presented September 15, 2024–January 19, 2025 at Chapman University. Orange, California.

– Elizabeth Hamilton, editor
– Freek Lomme/Set Margins',
publication advisor and publisher
– Willem Henri Lucas, publication designer
– Patrick J. Reed, managing editor

FULCRUM ARTS champions creative and critical thinkers at the intersection of art and science to provoke positive social change and contribute to a more vibrant and inclusive community.

This book was set in fonts Helvetica Neue, Times New Roman, and, especially for this project, Helvetica Vibrate Mono, a customized Helvetica Neue Bold Condensed, for the headlines.

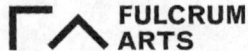

544 North Fair Oaks Avenue
Pasadena, CA 91103
www.fulcrumarts.org

Printed in Vilnius, Lithuania by BALTO print UAB on Multi Art Silk 150 gr. FSC (dust jacket), SORA Cream 70 gr. and Maxi Satin 100 gr. (interior), Rainbow 160 gr. black FSC (endsheets) in CMYK and gold PMS 871.

THE GUGGENHEIM GALLERY at Chapman University and Chapman University's Escalette Permanent Art Collection strive to engage with contemporary social issues, promote diversity and inclusion, and inspire critical thinking and creativity within the academic community and beyond.

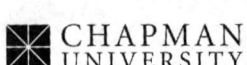

One University Drive
Orange, CA 92866
www.chapman.edu

Copublished by Fulcrum Arts and Set Margins'.

Creatively applying the cultural politics of content, form, and style whilst critically dissecting these to spark literacy, SET MARGINS' enforces the voices of marginalized cultural agendas.

First edition, 2024.

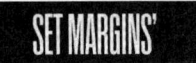

Set Margins'
Eindhoven, The Netherlands
www.setmargins.press

ISBN: ISBN 978-90-834498-0-7

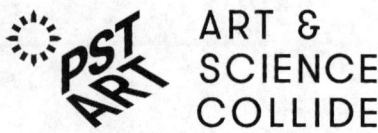

ART &
SCIENCE
COLLIDE

Presented by **Getty**

Energy Fields: Vibrations of the Pacific is made possible with the generous support of the Getty Foundation. Additional support provided by the Mike Kelley Foundation for the Arts and the National Endowment for the Arts. This project has also been assisted by the Australian Government through Creative Australia, its principal arts investment and advisory body. Further support has been provided, in part, by the LA County Department of Arts and Culture as part of Creative Recovery LA, an initiative funded by the American Rescue Plan. In-kind support provided by SOLARPUNKS and Heardrum.

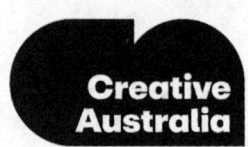

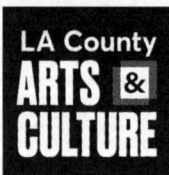

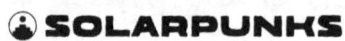

– Lawrence English, curator
– Robert Takahashi Novak, curator
– Marcus Herse, Guggenheim Gallery director
– Patrick J. Reed, assistant curator
– Marie Bland, development coordinator
– Jessica Bocinski, registrar
– Martin Carillo, sound designer
– Nicholas Cimiluca, web developer
– Emma Jacobson-Sive, public relations
– Jana Juhl, producer
– Karen Lofgren, grant writer
– Edward Patuto, development consultant
– Stella Peacock-Berardini, intern
– Sam Rowell, broadcast producer
– Tanya Rubbak, design
– Studio Sereno, fabrication and exhibition production
– Fiona Shen, education program manager
– Geneva Skeen, project coordinator
– Obed Soto Gonzalez, intern
– Sarah Stifler, public relations
– Jaspa Ureña, administrative assistant
– Scarlett Wang, digital media and marketing coordinator
– Tyler Wert, development researcher
– Holly Witham, accounting and finance
– Christopher Wormald, documentation

researchers
– Rana X. Adhikari
– Marcus Herse
– Vanessa Kwan
– W. Patrick McCray
– Enrique Rivera
– Fiona Shen
– Nina Tonga

advisors
– Joel Ferree
– Perrin Meyer
– Annea Lockwood
– Kyle Slabb
– Alex Wellerstein

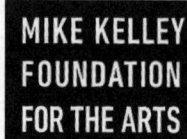

MIKE KELLEY
FOUNDATION
FOR THE ARTS

NATIONAL
ENDOWMENT for the ARTS
arts.gov

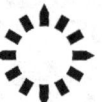

Energy Fields Vibrations of the Pacific, co-presented by Fulcrum Arts and Chapman University, participated in the PST ART Climate Impact Program, a groundbreaking integration of climate action, community building, and data reporting. Learn more at pst.art/climate.

ENERGY FIELDS II: VIBRATIONS OF THE PACIFIC

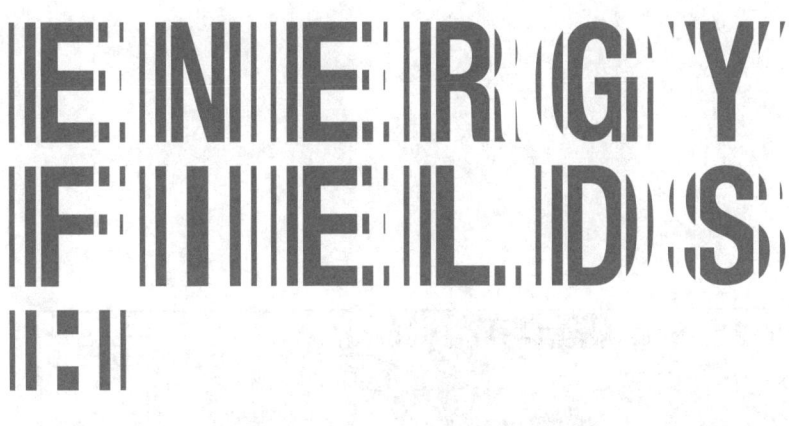

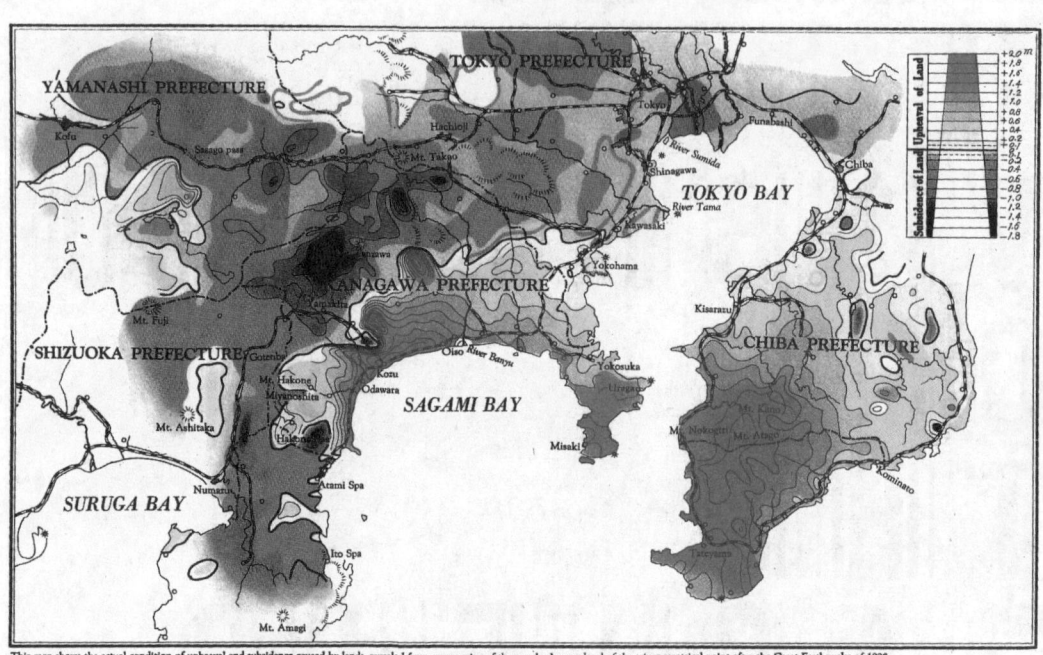

This map shows the actual condition of upheaval and subsidence caused by lands, revealed from resurveying of the standard water level of the trigonometrical point after the Great Earthquake of 1923.

Perpendicular changes to land caused by the
Great Kantō earthquake, Sagami Trough, Japan,
1923, from *The Reconstruction of Tokyo* (Tokyo:
Tokyo City, 1933).

ENERGY FIELDS II: VIBRATIONS OF THE PACIFIC

FULCRUM ARTS /
CHAPMAN UNIVERSITY

EDITORS
LAWRENCE ENGLISH
ROBERT TAKAHASHI NOVAK

AUTHORS
RANA X. ADHIKARI +
    AKANKSHA TIWARY
MARCUS HERSE
VANESSA KWAN
W. PATRICK McCRAY

PATRICK J. REED
ENRIQUE RIVERA
RACHEL SHEARER
FIONA SHEN

ARTISTS
WILLIAM BASINSKI
LAUREN BON AND
    THE METABOLIC STUDIO
JENEEN FREI NJOOTLI
ELLEN FULLMAN +
    THERESA WONG
DAVID HAINES +
    JOYCE HINTERDING
CHANNING HANSEN
VIRGINIA KATZ
BETHAN KELLOUGH

ANNEA LOCKWOOD
LEN LYE
ROSS MANNING
STEVE RODEN
MINORU SATO
RACHEL SHEARER
KYLE SLABB
AKIO SUZUKI
MALENA SZLAM
ALBA TRIANA
MO H. ZAREEI

PUBLISHER
SET MARGINS'

CONTENT

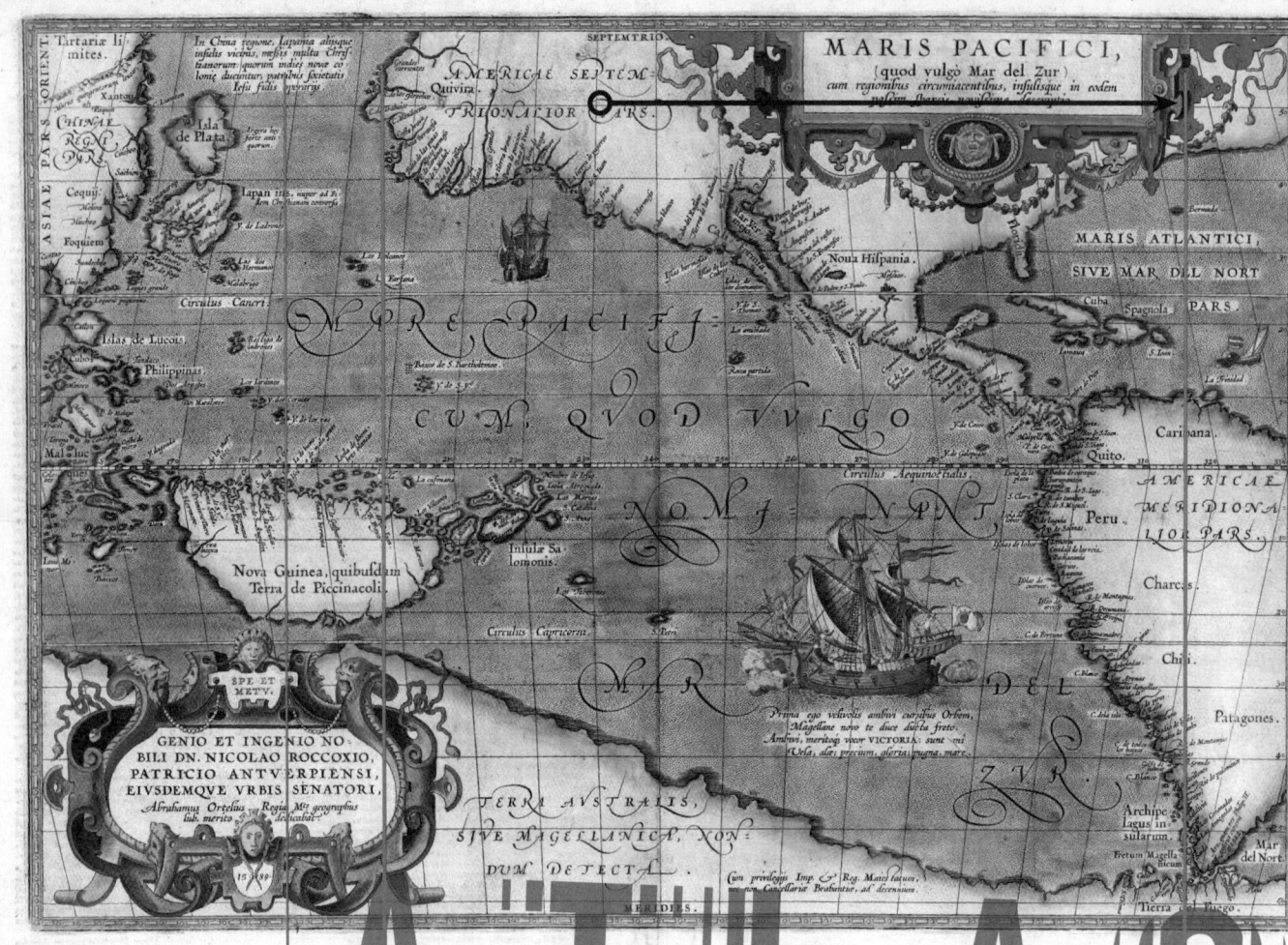

ATLAS OF WAVES

1. Rosie Alderson, "Humans Are Blind to 99.997% of Electro-magnetic Radiation," *Medium*, 22 September 2021, medium.com/everyday-science/humans-are-blind-to-99-9-of-electromagnetic-radiation-f15ec1215109.

2. Jens Blauert, *Spatial Hearing: The Psychophysics of Human Sound Localization* (Cambridge, Massachusetts: The MIT Press, 1996), 2.

3. Adam Weisser, "Treatise on Hearing:

Maris Pacifici (quod vulgo Mar del Zur), the first printed map of the Pacific, published in Antwerp, 1589. Photograph courtesy the Barry Lawrence Ruderman Map Collection, David Rumsey Map Center, Stanford Libraries.

ENERGY FIELDS: VIBRATIONS OF THE PACIFIC

by LAWRENCE ENGLISH

Almost everything around us—and even inside us—is in a state of vibration. From the sub-atomic to the tangible, the oceanic, the planetary, and even the galactic, vibration operates on almost everything in ways we have struggled to imagine, let alone understand. It is a constant flux of pulse and motion that invites imagination, research, and embodiment with equal vigor.

When an object comes out of a state of equilibrium, it comes into vibration; think of a puddle disturbed by a droplet of water, or a bridge quivering because of heavy traffic. Many vibrations happen on scales we cannot comprehend. Consider our two predominant senses: sight and hearing. The narrow range of information we glean from them is a mere fraction of what is unfolding in any given moment. Just 0.0035% of the electromagnetic spectrum is visible to the human eye,[1] making ultraviolet light, for example, undetectable. Our ears are equally limited, offering an impression of sound that proximately exists between twenty hertz and twenty kilohertz,[2] such that detecting the infrasonic murmurings of earthquakes and the ultrasonic chattering of bats remains outside our sensory capacities.

When an object vibrates, it causes the space around it to shift; we call these shifts waves. Picture a small pool: the water is still, but you decide to throw a stone into it. The stone hits the water, displacing it, and the disturbed mass pushes outward and then back again over the stone as the stone sinks. This causes the water to ripple outward from the stone in waves. Waves are the temporary result of specific vibrations, and they operate on bodies, objects, and even space, communicating about vibrational events whether we were there to witness them or not.[3] The degree to which we are aware of vibrations and their resulting waves is highly variable.[4] While our experience of the world is shaped by these encounters, the limitations of our anatomy make our observation of them finite and conditional.

Our capacity to sense and comprehend the vibrational fields that surround us has been a robust area of research over the past few centuries. Increasingly, this research has come to recognize the material consequences of vibration thanks to technological developments that allow vibrational events to be reproduced and observed. For example, Vladimir Gavreau, a French scientist interested in the effect of sound on the body, described some of the physiological implications of vibration experienced while studying infrasonic wavefields during the 1960s.[5] While Gavreau couldn't necessarily hear the sounds being produced, he could feel them and even see them as objects in his laboratory began to shake. Members of his staff also described significant physical reactions such as headaches and pain in the ears, which we now understand as the effects of exposure to intense infrasonic sound.

Over the past century, technological, scientific, and philosophical advancements have allowed us opportunities to ponder, test, and reveal a multiplicity of theories and actualities. Acoustic levitation, a process by which objects are held in that air by focused sound waves, existed as a speculative promise in the early twentieth century but is now possible via Phased Array Ultrasonic Transducers.[6]

The Temporal Auditory Imaging Theory Inspired by Optics and Communication," *arXiv: Neurons and Cognition* (November 2021), arxiv.org/pdf/2111.04338.pdf.

4. See Christoph Cox, "Beyond Representation and Signification: Toward a Sonic Materialism," *Journal of Visual Culture* 10, no. 2 (August 2011): 145–61; Walter S. Gershon, "Vibrational Affect: Sound Theory and Practice in Qualitative Research," *Cultural Studies, Critical Methodologies* 13, no. 4 (August 2013): 257–62; and Steve Goodman, *Sonic Warfare: Sound, Affect, and the Ecology of Fear* (Cambridge, Massachusetts: The MIT Press, 2012).

5. Mark Pilkington, "Extreme Noise Terror," *The Guardian*, 18 June 2003, theguardian.com/education/2003/jun/19/research.highereducation2.

6. Marco A. B. Andrade, Asier Marzo, and Julio C. Adamowski, "Acoustic Levitation in Mid-Air: Recent Advances, Challenges, and Future Perspectives," *Applied Physics Letters* 116, no. 25 (June 2020).

INTRODUCTION ATLAS OF WAVES

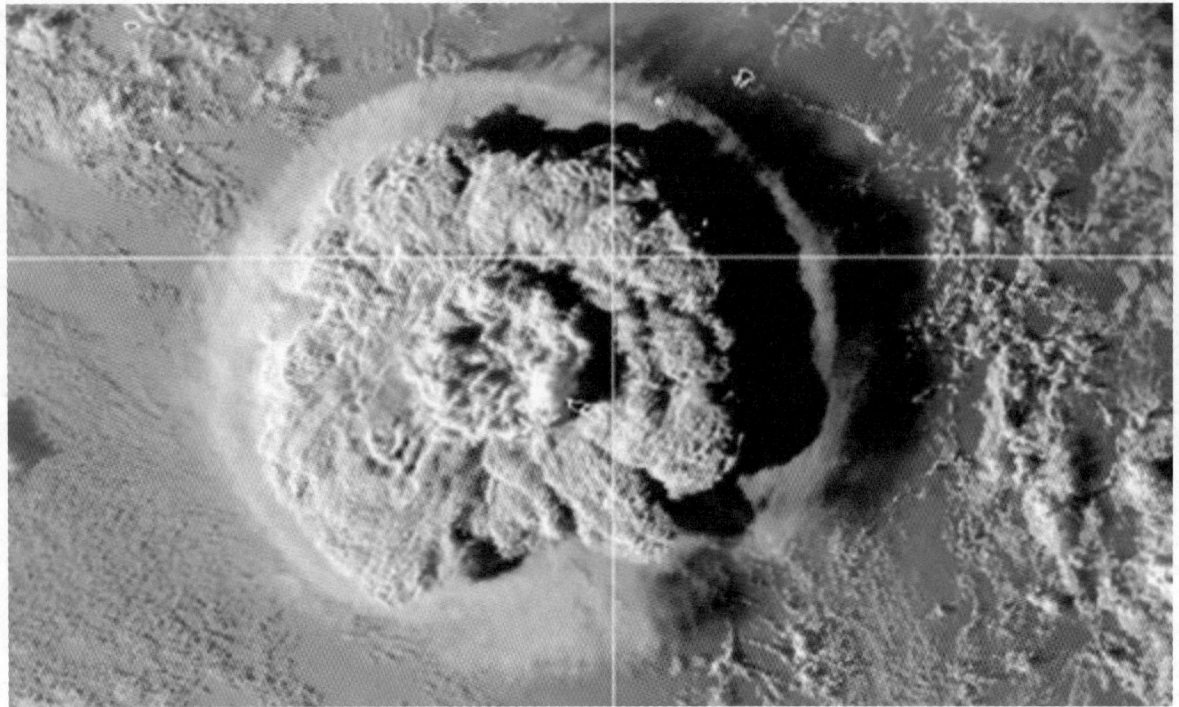

Eruption of the Hunga Tonga–Hunga Haʻapai submarine volcano, Tonga, South Pacific, 2022. Photograph courtesy Reuters.

7. Don Ihde, *Listening and Voice: Phenomenologies of Sound*, 2nd ed. (Albany, New York: State University of New York Press, 2007).

8. Lawrence English, "Relational Listening: A Politics of Perception," *Contemporary Music Review* 36, no. 3 (2017): 127–42.

In the realm between the known and the unknown, the sensed and the sensing, an aspirational ontological position has emerged—an ontology of vibration that seeks to approach, recognize, and interpret the world through the flow of its energies. The promise of this vibrational ontology is at the core of *Energy Fields: Vibrations of the Pacific*. The exhibition prompts a consideration of the material and immaterial vibrations that surround us. Furthermore, it encourages us to be celebratory and aspirational in our interrogation of ideas and experiences that may exist outside the realm of everyday comprehension, helping us test the limits of what we can propose, detect, and catalog as fact.

TOWARD VIBRATIONAL KNOWING

Vibration is an invitation, a provocation even, used by artists and scientists alike to push beyond what we already know and can describe. Imagine yourself in a room: though you can look around the space, there are limits to the scope of your observation. Light is visible, but its intensity is affected by material conditions such as windows or walls. Sounds are audible, some of which are originating from outside the room and reveal information about the world beyond what you can see. Touch also might reveal something about conditions outside the room—the movement of a window pane might reveal a howling wind. Low-frequency sounds can also be felt, leading some to characterize the body as an extension of the ear.[7] And, with you in the room are tools and technologies—a radio, a pair of infrared goggles—that permit you to detect vibrations otherwise beyond your senses. Such devices may push your capacity for knowing into a zone of entanglement that is still being understood and, in many cases, imagined.[8]

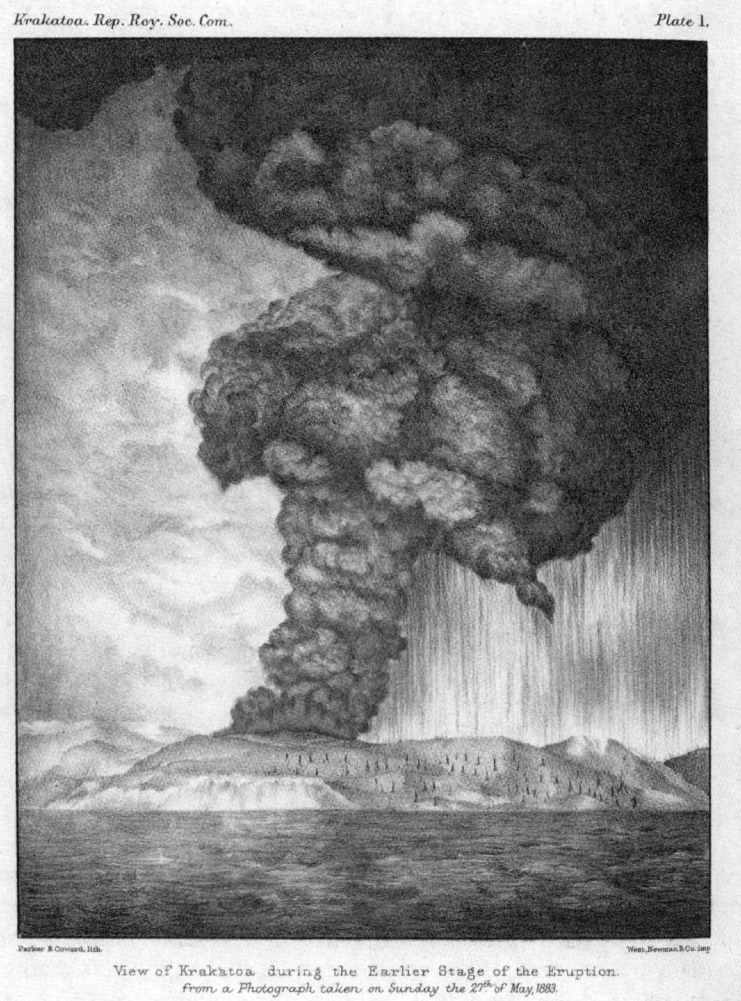

Krakatoa. Rep. Roy. Soc. Com.   Plate 1.

Farber & Coward, lith.   West, Newman & Co. imp

View of Krakatoa during the Earlier Stage of the Eruption.
from a Photograph taken on Sunday the 27th of May, 1883.

Lithograph depicting the 1883 eruption of the Krakatoa caldera in Indonesia, from *The Eruption of Krakatoa, and Subsequent Phenomena. Report of the Krakatoa Committee of the Royal Society* (London, Trubner & Co., 1888).

9. Jorge L. Cervantes-Cota, Salvdor Galindo-Uribarri, and George F. Smoot, "A Brief History of Gravitational Waves," *Universe* 2, no. 3 (September 2016): 22.

Our current understanding of gravitational waves, for example, traces a long line from theoretical proposition to observable phenomenon. Gravitational waves are the remnants of enormous energetic ruptures—exploding stars or colliding black holes—that change the curvature of spacetime and ripple across the galaxies. Originally theorized by Albert Einstein in 1916 as part of his Theory of General Relativity, these waves were first detected a decade ago, when the Laser Interferometer Gravitational-Wave Observatory (LIGO) sensed minuscule undulations in the fabric of spacetime caused by two black holes colliding 1.3 billion light years away.[9] The amount of fluctuation these waves caused was 10,000 times smaller than the nucleus of an atom, a testament to LIGO's sophistication.

Discoveries such as this frame up why we should concern ourselves with vibrations. These energetic forces prompt us toward an understanding of our world and ourselves that extends beyond our phys-

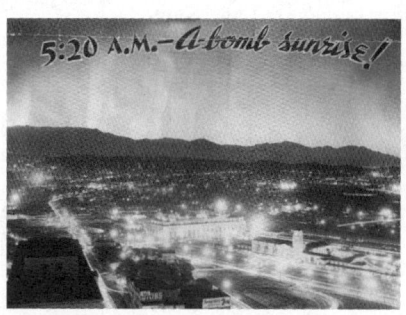

How the largest of the Nevada atomic blasts looked from Los Angeles, 1955, Herald Examiner Collection. Photograph courtesy of Los Angeles Public Library.

INTRODUCTION ATLAS OF WAVES

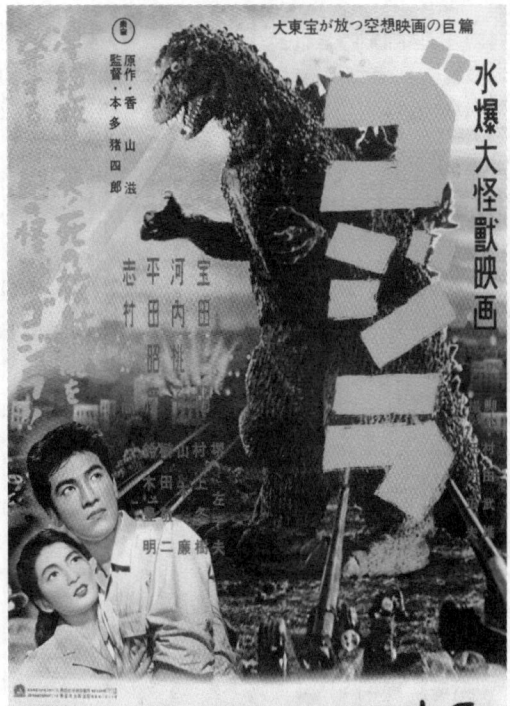

Poster advertising *Godzilla*, dir. Ishirō Honda (1954; Tokyo: Toho Co., Inc).

10. David M. Boore, "The Richter Scale: Its Development and Use for Determining Earthquak Source Parameters," *Tectonophysics* 166, no. 1–3 (September 1989): 1–14.

11. Corwin J. Wright, Neil P. Hindley, M. Joan Alexander, et al., "Surface-to-Space Atmospheric Waves from Hunga Tonga–Hunga Ha'apai Eruption," *Nature* 609, no. 7928 (June 2022): 741–46.

12. Kevin Hamilton, "Tonga Eruption Was So Intense, It Caused the Atmosphere to Ring Like a Bell," *The Conversation*, 23 January 2022, theconversation.com/tonga-eruption-was-so-intense-it-caused-the-atmosphere-to-ring-like-a-bell-175311.

13. Takashi Tonegawa and Yoshio Fukao, "Wave Propagation of Meteotsunamis and Generation of Free Tsunamis in the Sloping Area of the Japan Trench for the 2022 Hunga–Tonga Volcanic Eruption," *Earth, Planets and Space* 74, no. 1 (November 2022): 159.

ical senses—what we already know or can observe. To engage with a vibrational ontology is to strive toward that which lies at the edges of knowing by encouraging deeper and more embedded theoretical and methodological approaches across disciplines. A vibrational ontology asks us to open ourselves, as researchers, as creators, and as humans to the presence and potential of the energies around us, and it reminds us that knowledge and the ways of being that follow it are in a constantly fitful state, tacking back and forth between the theoretical and the experiential. And here, at this nexus of aspiration and discovery, is the interplay of art and science. The desire to pierce into the darkness of uncertainty in the hope of discerning new ways of being, knowing, and sensing is at the very heart of both creative pursuits.

WAVES AND RUPTURES

As its title suggests, *Energy Fields: Vibrations of the Pacific* is concerned with a specific geography. The Pacific encompasses a multitude of landscapes, cultures, languages, and communities. It is home to many peoples who trace long histories, including the oldest continuous culture in the world, that maintained by the Aboriginal nations of Australia. In 1589, the first printed map of the Pacific, *Maris Pacifici*, was made in Antwerp. Incomplete and inaccurate as it is, it speaks to a recognition of this geographic region as linked, with the ocean acting as a tether, a connective tissue between dynamic, muscular landmasses. That map was the first of many that would portray the region's geologic, topographic, oceanographic, geographic, social, economic, and political conditions. It allowed others to conceive of a zone where the connections might be aquatic and tidal rather than terrestrial.

The Pacific is also home to some of the most energetic places in the world. Underneath the ocean sit tectonic plates that are among the most volatile and active on earth, having produced some of the most consequential vibrations on the planet. Over the past decades, we have witnessed some of their enormous energetic outpourings. In 2011, the Tōhoku earthquake and tsunami in Japan, also known as the Great Eastern Earthquake, was one of that nation's most devastating natural disasters. The quake registered a magnitude of 9.1 on the Richter scale,[10] a measurement system devised by Charles F. Richter and Beno Gutenberg in 1935 to quantify the amount of energy released by seismic events. The results were catastrophic and are still being felt;

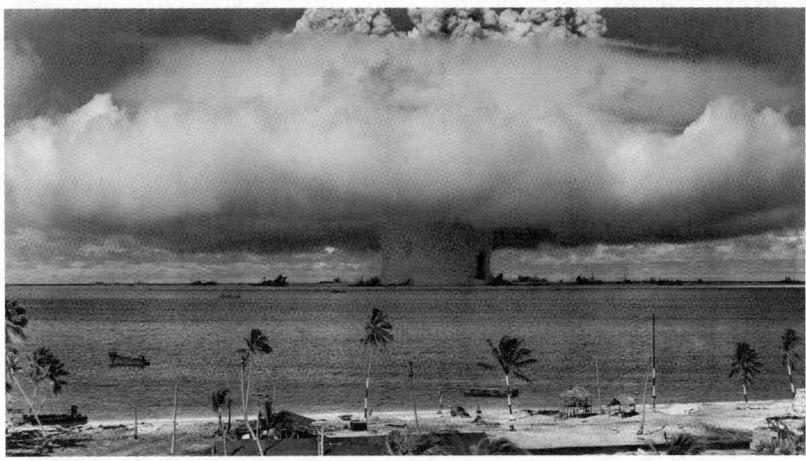

Operation Crossroads, Bikini Atoll, Marshall Islands, 1946.

14. Paul Boyer, *By the Bomb's Early Light: American Thought and Culture at the Dawn of the Atomic Age* (Chapel Hill: University of North Carolina Press, 2005).

15. Mike Ladd, "The Lesser-Known History of the Maralinga Nuclear Tests—and What It's Like to Stand at Ground Zero," *ABC News Australia*, 23 March 2020, abc.net.au/news/2020-03-24/maralinga-nuclear-tests-ground-zero-lesser-known-history/11882608.

16. David Ropeik, "How the Unlucky Lucky Dragon Birthed an Era of Nuclear Fear," *The Bulletin of Atomic Scientists*, 28 February 2018, thebulletin.org/2018/02/how-the-unlucky-lucky-dragon-birthed-an-era-of-nuclear-fear/.

towns such as Ishinomaki were not only partially destroyed, but their geography was permanently altered. In Fukushima, damage to a nuclear power plant released radioactive material into the air and water, with effects that will linger for decades.

In January 2022, the Hunga Tonga-Hunga Ha'apai volcanic eruption near Tonga produced so much energy it created a pressure wave, visible from space, that emanated from the blast point.[11] This echoed the 1883 Krakatoa eruption in Indonesia, a blast so strong the sound was audible across the Pacific.[12] Both Krakatoa and Hunga Tonga-Hunga Ha'apai released so much force they rang the Earth like a bell: they generated Lamb waves (waves that travel through solid material) that pulsed back and forth across the globe three times, and acoustic gravity waves that circled the earth at least six times before dissipating. Traveling at 1,125 feet per second, roughly the speed of sound, the acoustic gravity waves were strong enough to generate massive changes in air pressure, or meteorological tsunamis, across Europe.[13]

The Pacific has also been a place for human-made energy and vibrations—specifically, atomic energy. During the mid-twentieth century, it was the site of nuclear bomb detonations. In a vulgar display of power, the United States dropped bombs on the Japanese cities Hiroshima (the detonation site for *Little Boy*, an atomic bomb) and Nagasaki (the detonation site for *Fat Man*, a plutonium bomb), the first and only use of nuclear weapons during wartime.[14] Rather than curbing an emergent fascination with nuclear warfare, these bombings ushered in a period of exhaustive research on the energetic potential of atomic weaponry. Atomic weapons testing also displaced many Marshall Islands communities and had significant health impacts on subsequent generations of Marshallese. Similarly, the British government's nuclear arms testing in Australia at Maralinga led to the long-term harm and displacement of the Maralinga Tjarutja people.[15] In 1954, a thermonuclear test at Bikini Atoll resulted in fallout that reached a Japanese fishing boat; the crew suffered significant radiation sickness.[16] This incident was sharply felt by the Japanese, whose encounters with the horror of nuclear blasts were still firmly in the collective consciousness. Flying

17. Peter H. Brothers, "Japan's Nuclear Nightmare: How the Bomb Became a Beast Called 'Godzilla,'" *Cinéaste* 36, no. 3 (Summer 2011): 36–40.

18. Alfred Korzybski, *Science and Sanity: An Introduction to Non-Aristotelian Systems and General Semantics* (New York: Institute of General Semantics, 1958), 58.

19. Rachel Shearer, "Te Oro o te Ao: The Resounding of the World," PhD diss. (Auckland, New Zealand: Auckland University of Technology, 2018.)

over Bikini Atoll that year, Japanese film producer Tomoyuki Tanaka conjured up the story that would birth Godzilla (Gojira in Japan), one of the Pacific's most significant pop-cultural manifestations of nuclear tragedy.[17]

THE MAP IS NOT THE TERRITORY

Energy Fields: Vibrations of the Pacific meshes together a wide array of themes, sources, and perspectives from across art, politics, social history, geography, and science. It acts as a relational atlas, a thematic volume that seeks to be porous, to embrace all the possibilities for sensory experience, and to recognize the complexity of exploring concepts and terrains through multiple methodologies in the same moment. The projects and artworks the exhibition presents are portals through which multiple understandings and inquiries can be realized at once.

The relations explored in the exhibition are multifarious and at times even tangential. They chart connections between phenomena, culture, knowledge systems, and place, and are offered as a provocation towards restlessness, towards a desire to test for the boundaries of practice, of understanding, and of application. As Alfred Korzybski summarized, the "map is not the territory."[18] What it is, however, is a launching point from which new perspectives might emerge. In the same way *Maris Pacifici* invited readers to consider the Pacific as connective and interrelated, this exhibition hopes to embrace that appreciation for curiosity and the wonder that it brings. Using vibration as a guiding principle and ontological device, the collection of artworks, texts, artist statements, interviews, and documents that make up *Energy Fields: Vibrations of the Pacific* recognizes the aspirational work that sits at the heart of scientific or creative research is never complete but rather is a process of knowledge sedimentation and stratification across time.

This exhibition also seeks to raise critical questions about how we consider vibration as a force for change—a catalyst for new understandings of ourselves, our planet, and our universe. It encourages a deepening of research and responses (creative or otherwise), and it aspires to embrace the excessive energies that operate on and through us, moment to moment. It recognizes the body as a first point of contact, the receiver of sensory input that shapes our most fundamental understanding of the world. But it also pushes outward, embracing ways of knowing informed not just by scientific theory and technological advancements, but also by the knowledge of First Nations people and Indigenous communities across the region, many of which have developed their own ontologies of energy, vibration, and sense-making.[19] Like the Pacific itself, this exhibition is porous and fluid, absorbing a multiplicity of readings and creating a tidal flow of ideas and expression.

We hope artists, researchers, and scientists are inspired to embrace vibration as a dynamic source for exploring all manner of unexpected and hidden phenomena in our midst. <

by RACHEL SHEARER

A WHAKA OF SHIMMI

Rachel Shearer, still from *Whakapapa of Shimmers*, 2024.

Kia hora te marino
May peace be widespread
Kia whakapapa pounamu te moana
May the sea be like greenstone
Kia tere te kārohirohi
May the shimmer of light
i mua i tō huarahi
guide you on your way.[1]

t is midsummer in the north of Aotearoa (New Zealand). In a small stand of trees, the shimmering of the season intensifies. Shimmering heat, shimmering sound. It's early afternoon, and cicadas' mating communications register on the decibel meter app on my iPhone as "92db—alarms/power tools." The volume increases as the weather gets hotter. *Kihikihi, kikihi, tātarakihi, terakihi*[2] —

the males chant their own names, forty-plus species of cicadas chorusing around these islands.

Heated air rising from the earth appears to shimmer—light refracting off the ground passes back up through atmospheres of different thermal densities. This shimmer effect is described in the oral traditions of some *iwi* (extended kinship groups) as *Te Haka-a-Tānerore* (the Dance of Tānerore), the metaphorical son of the sun and summer.[3] The Dance of Tānerore is invoked in the Māori performing arts to excite one's energies by attuning body and mind to his energetic shimmer. *Ko koutou mā ērā i poipoi i te kārohirohi o te haka a Tāne-rore* (You are necessary to nurture the shimmer of Tānerore's *haka*).[4]

Tānerore's companion, the feminine glistening Hineruhi, comes at dawn. The sparkle of light reflected in the morning dew is her dance—he shimmers, she sparkles. A woman who excels in performance would be complimented with *Ko Hineruhi koe, nāna i tū te ata*

---

1. As explained by Taituwha King (Waikato), this *whakatauākī* (proverb) was written by Rangawhenua (Ngāti Pahere, Ngāti Maniapoto) for his own people and for a specific context. It has since been widely adapted and used across Aotearoa (New Zealand), often for those about to embark on a journey or an important life passage.

2. Some of the onomatopoeic names various *iwi* (extended kinship groups) have given the cicada. *Te Aka Māori Dictionary*, s.v. "Cicada," maoridictionary.co.nz/, accessed May 2023.

3. Te Ahukaramū Charles Royal, "Te Whare Tapere: Towards a Model for Māori Performance Art," PhD diss. (Victoria University of Wellington, New Zealand, 1998), 132.

4. Ngāti Kahungunu invitation for the biennial 2017 Te Matatini Kapa Haka Festival.

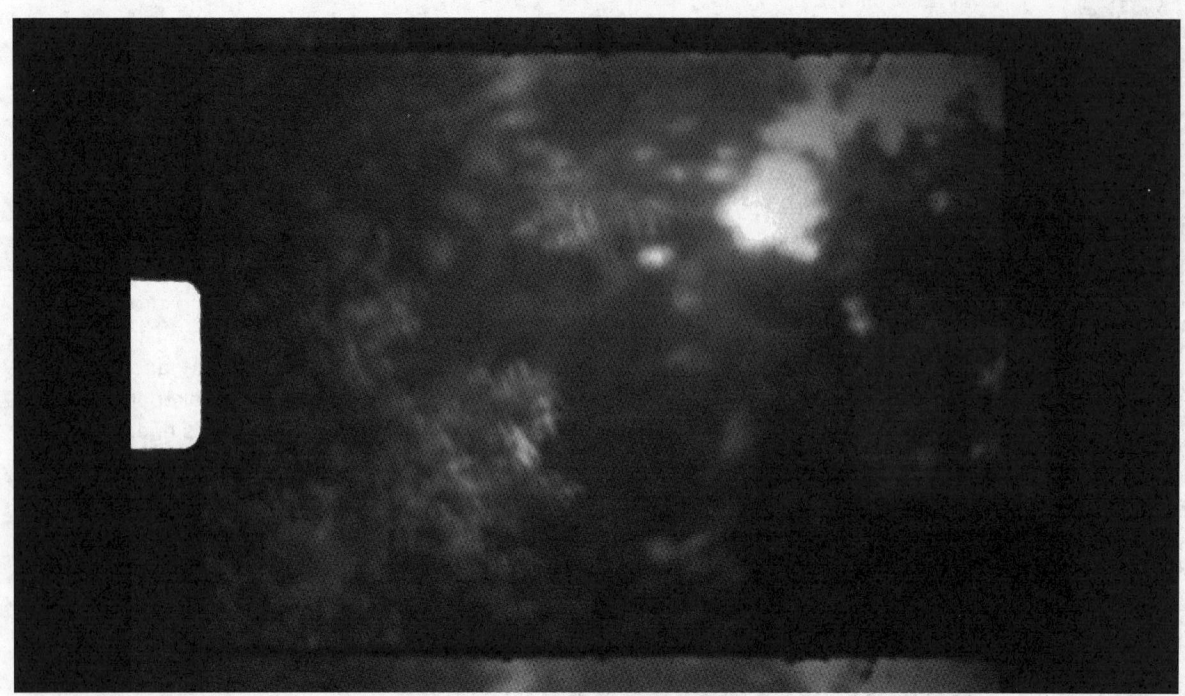

Rachel Shearer, *wiriwiri*, still from *Whakapapa of Shimmers*, 2024.

hāpara (You are like Hineruhi, the woman who ushers in the dawn).[5] Te Ahukaramū Charles Royal (Ngāti Whanaunga, Ngāti Tamaterā) explains that both the Tānerore and Hineruhi traditions give physical manifestation to environmental phenomena as performed by humans. "In these characters of Tānerore and Hineruhi, male and female dance finds both their mythological bases and their finest examples by which all standards are set."[6]

A feature of Māori performance traditions is the quivering, trembling, shimmering hand gesture, the *wiriwiri* (to tremble, quiver, shake), an acknowledgement of Tānerore and his dance.[7] This gesture evokes the idea of vibration at the core of being, the shimmer of light on water, the rustle of leaves in the trees, the heat waves of Tānerore rising from the ground reverberating through bodies, emotion, memories, imagination, space, time, and realms.

"What pattern connects the crab to the lobster and the orchid to the primrose and all the four of them to me? And me to you?," asked systems-thinking innovator Gregory Bateson.[8] He saw an interconnected universe as pattern and fabric, texture and weave. *Whakapapa* is the name of the interconnecting patterns of physical, elemental, and spiritual forces in *te ao Māori* (the Māori world). *Whakapapa* is the web that locates things in space and time and their relationship to each other. Through its multiple contexts, *whakapapa* can mean to layer (flattish) things; it refers to genealogy, it relates to the practices of the retelling of relationship connections and the stories embedded within those connections. In this way, it is a system of organizing knowledge and structuring a worldview. Throughout these relations is the energetic substance called *mauri*, a life force that imbues the material world and binds it to metaphysical realms.[9] The deification of natural phenomena in the Māori world tells the grand story of interconnection of humans to the earth. Keri Opai (Te Atiawa, Ngāti Ruanui) describes *atua* (such as the Tānerore) as an *a-tua*. The grammar shift changes the translation to mean something that is "beyond"— beyond what we can control or know.[10] The *atua* Ranginui, his name often translated as the Sky Father, is described by Carl Mika (Tūhourangi) as the Grand Resonance, a term he claims as imperfect but nonetheless a gesture towards trying to communicate in English the depth and vastness of the concept of Ranginui.[11] In the same way, Papatūānuku, the Earth Mother, is the ground of our thinking, our source and sustenance. We are genealogical extensions of these *atua*. *Ko Ranginui e tū iho nei, Papatūānuku e tāko-to nei* (Ranginui stands above, Papatūānuku lies below).[12]

Regarding the *whakapapa* of Tānerore, while different *iwi* and regions have different stories, a common version is that Ranginui and Papatūānuku's eldest son, Uruten-gangana, *atua* of light, and Hineteāhuru, *atua* of warmth, are the parents of Tamanuiterā, the sun.[13] Tamanuiterā and Hineraumati, the summer, are the parents of Tānerore. This *whakapapa* can also be understood as empirical data, observations woven into stories that impart multiple levels of information at once—relationships of light and heat and moisture as Tānerore and Hineruhi, the alluring aesthetics of the shimmer effect. Science might tell you that the shimmer is due to a change in air's index of refraction, which causes light to travel along curved paths, creating fluctuations in intensity and color. What this doesn't describe is the sense of joy that entertainment brings, exemplified by the shimmers and sparkle of Tānerore and Hineruhi as expressed in this *whakataukī* (proverb): *Kia kawea tātou e te rēhia* (Let us be taken by the spirit of joy, of entertainment).[14]

The shimmer as an aesthetic affect and an experience of joy is also located in art anthropologist Howard Morphy's 1989 discussion of *biry'un* (brilliance or shimmering) in the paintings of the First Nation Australian people the Yolgnu.[15] Between 1973 and 1976, Morphy was taught by the Yolgnu about the significance of the cross-hatching effect in their art making, referred to as *bir'yun*. He was told that *bir'yun* in its everyday use refers to sources and refractions of light such as the sparkling of bubbling fresh water. It is this quality of brilliance that is associated with beauty, feelings of lightness, joy, and ancestral power (though in some contexts, this power can be dangerous, so care is important).[16] These paintings re-create ancestral designs, produced in ceremony as part of the re-creation of ancestral events, and are also a demonstration of rights held by a clan in

5. Royal, "Te Whare Tapere," 183.

6. Ibid., 136.

7. Rachel Ka'ai-Mahuta, "The Genesis of Waiata and Haka," in *Kia Rōnaki: The Māori Performing Arts*, ed. John Moorfield, Tania Ka'ai, and Ka'ai-Mahuta (Auckland: Pearson New Zealand, 2013), 5.

8. Gregory Bateson, *Mind and Nature: A Necessary Unity* (New York: E. P. Dutton, 1979), 8.

9. Māori Marsden, *The Woven Universe: Selected Writings of Rev. Māori Marsden*, ed. Te Ahukaramū Charles Royal (Otaki, New Zealand: Estate of Rev. Māori Marsden, 2003).

10. Keri Opai, *Tikanga: An Introduction to Te Ao Māori* (Auckland: Upstart Press, 2021), 42.

11. Carl Mika, "The Interconnection of Things in an Era of Colonisation: Māori and Education," YouTube, 1:30:16, 24 February 2021, youtube.com/watch?v=yMXQWIku68I, accessed June 2023.

12. This phrase is often heard in *karakia* (incantations, applications to *atua*, prayer).

13. Rangi Matamua, "Living by the Stars Tamanuiterā EP1—Tamanuiterā," YouTube, 18 April 2023, 4:45, youtube.com watch?v=KHtYEubv4YE, accessed June 2023.

14. Royal, "Te Whare Tapere," 184.

15. Howard Morphy, "From Dull to Brilliant: The Aesthetics of Spiritual Power Among the Yolngu," *Man* 24, no. 1 (March 1989): 21–40.

16. Ibid., 30.

mardayin (body-painting ritual) and land.[17] As ancestral designs that originated through actions in the ancestral past, the *biry'un* effect is interpreted as a manifestation of ancestral power emanating from the ancestral past.

The shimmer as ancestral power is also discussed by ecologist Deborah Bird Rose through her work with another First Nation Australian people, the Yarralin. Concerned for Anthropocene-era-affected flying foxes in Australia, she lived with the Yarralin to learn about their relationship with this species. There, flying foxes are woven into the identity of the community. Once Rose was claimed by one of the senior women of the community, a flying-fox matriarch, flying foxes became her kin. Through her time as Yarralin flying-fox kin, Rose came to understand the shimmer as connoting life as a flow of ancestral power across the biosphere, where everything is communication—trees call to flying foxes via their nectar, and flying foxes call for rain.[18] Eucalyptus flowers calling out, "Yes!" The flying foxes responding "Yes!" to the call of the flowers. "Yes! Becoming ancestral across generations and boundaries. Becoming part of the great shimmer, from outside to inside, and from inside back out again."[19] Rose describes creating patterns of communication through Yarralin ceremony, where through the "interweaving patterns" of dance and sound and the earth and the sky, there is a back and forth, an oscillation, between call and response.[20]

The world perceives us as we perceive the world: there is an exchange. The image of an oscillating call and response is present in biologist Lynn Margulis's study of proprioception of the earth. She explains that "sensitivity, awareness, and responses of plants, protoctists, fungi, bacteria and animals each in its local environment, constitute the repeating pattern that ultimately underlies global sensitivity and the response of Gaia 'herself.'"[21] In this context, Margulis is referring to an idea of the Earth/Gaia as a closed system, where an assumed boundary of the interior and exterior of the system prevents matter flowing in or out. However, it does allow the transfer of energy in and out of the system. Depending on whether your understanding of energy

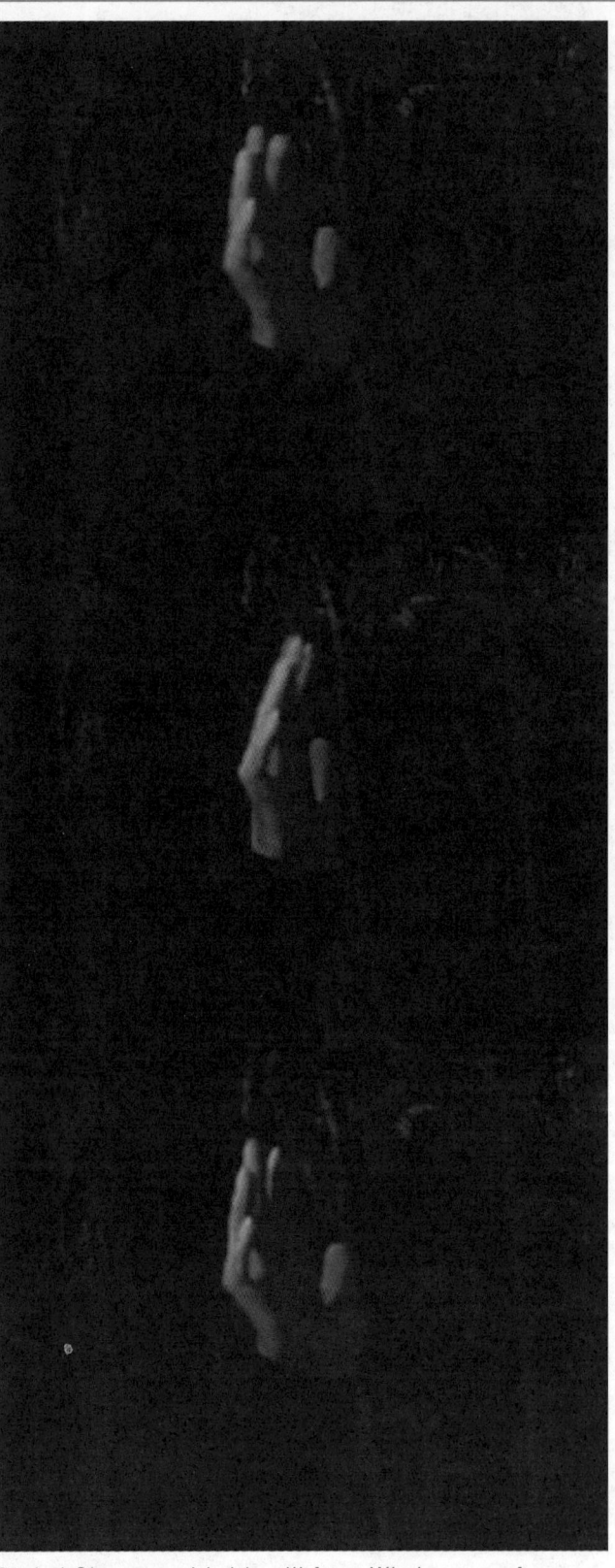

Rachel Shearer, *wiriwiri*, still from *Whakapapa of Shimmers*, 2024.

includes those that defy measurement (such as spiritual forces), then we could consider some correlation with Ngā Puhi philosopher Māori Marsden's description of a universe where spiritual and physical worlds are inter-related and can be affected or modified by each other. This is a worldview where the natural order was not a closed system, it could be infiltrated and interpenetrated by the higher order of spirit.[22] This system is interwoven with *mauri* and *mana*, amongst a cluster of other energetic concepts that intermingle across the human and non-human worlds.[23] While the oscillations of call and response within *te ao Māori* are apparent in the art of the *karanga*, the women's call and response performed during formal rituals of encounter,[24] here I want to consider oscillations of call and response with the world through an exchange of *mauri* guided by the recognition of *mana* in others. Mason Durie (Rangitāne, Ngāti Rau-kawa) explains, "Within a Māori ontological frame, all beings and objects are experienced as having *mana*, a form of presence and authority, and a 'vigour, impetus, and potentiality' called *mauri*."[25] *Mauri* is described in the Te Aka Māori Dictionary as "life principle, life force, vital essence… the essential quality and vitality of a being or entity."[26] As Te Kawehau Hoskins (Ngāti Hau) and Alison Jones explain, in the Māori world, it is the relationship between things that is the most important. It is the energetic exchange rather than the "thing" itself, that is ontologically privileged.[27] Maintaining *mauri* is an ethical practice, informing interactions with all the various entities and ecologies that we share existence with.

Stars, objects on the horizon, lights of a city all seem to shimmer when observed at a distance. It is especially noticeable when viewed through a turbulent atmosphere; water surfaces, hot pavement, wind, all causing fluctuations of what we see. Information on, off: 1, 0, 1, 0, 1, 0. While it would be poetic to imagine digital data flows as shimmers, there is the matter of ecological impact due to the energy demands of the physical infrastructure that digital data requires. The hardwired networks of fiber-optic cables, servers, routers, and the spaces, technologies, and nonregenerative materials that make them, house them,

17. Ibid., 26.

18. Deborah Bird Rose, *Shimmer: Flying Fox Exuberance in Worlds of Peril* (Edinburgh: Edinburgh University Press, 2022), 144.

19. Ibid., 228–30.

20. Ibid., 85.

21. Lynn Margulis, in Bruce Clarke, *Gaian Systems: Lynn Margulis, Neocybernetics and the End of the Anthropocene* (Minneapolis: University of Minnesota Press, 2020), 12.

22. Marsden, *The Woven Universe: Selected Writings of Rev. Māori Marsden*, 20.

23. If we consider vitalism as a theory that understands that the phenomena of life is not just chemical or physical, a Māori vitalism can be located in the belief that an immanent life force "imbues and animates all forms and things of the cosmos." The life forces known as *mauri* and *mana* exist within an intricate relationship of metaphysical concepts—*tapu*, *wairua*, and *hau*—that together inform the expression of a Māori vitalism. See Mānuka Hēnare, "Tapu, Mana, Mauri, Hau, Wairua: A Māori Philosophy of Vitalism and Cosmos," in *Indigenous Traditions and Ecology: The Interbeing of Cosmology and Community*, ed. John Grim (Cambridge, Massachusetts: Center for the Study of World Religions, Harvard University, 2001), 197–21.

24. While simultaneously opening up portals between space, time, and realms, but that is another story.

25. Mason Durie, in Te Kawehau Hoskins and Alison Jones, "Non-human Others and Kaupapa Māori Research," in *Critical Conversations in Kaupapa Māori* (Wellington: Huia Publishers, 2017), 52.

26. *Te Aka Māori Dictionary*, s.v. "Mauri," maoridictionary.co.nz/, accessed July 2023.

27. Hoskins and Jones, "Non-Human Others and Kaupapa Māori Research," 53.

and cool them.[28] The material requirements to create and transmit technological shimmers are an uneasy detail in an engagement with the digital arts.

From energetic exchanges and transformations, to information as pattern and metaphor that help construct the stories we tell, this is a *whakapapa* of shimmers. People interpret the mysteries according to their own worldview. My understanding of *te ao Māori* is translated across colonized thought patterns; decolonize, unweave, indigenise, reweave, new patterns emerge. Beyond physics and the impact of the hardware required to gen-

erate technological shimmers, the symbolic dimensions of this phenomenon remain. The shimmer becomes a metaphor for the moment, the happening, the potential that results from energetic exchanges between elements and things and species. The shimmer associates with metaphysical energies—ancestral, spiritual, and the energetic becomings of aesthetic affect, movement, and joy. Life!

May the shimmer guide you on your way. ◂

28. Kyle Devine, *Decomposed: The Political Ecology of Music* (Cambridge, Massachusetts: The MIT Press, 2019), 129–64.

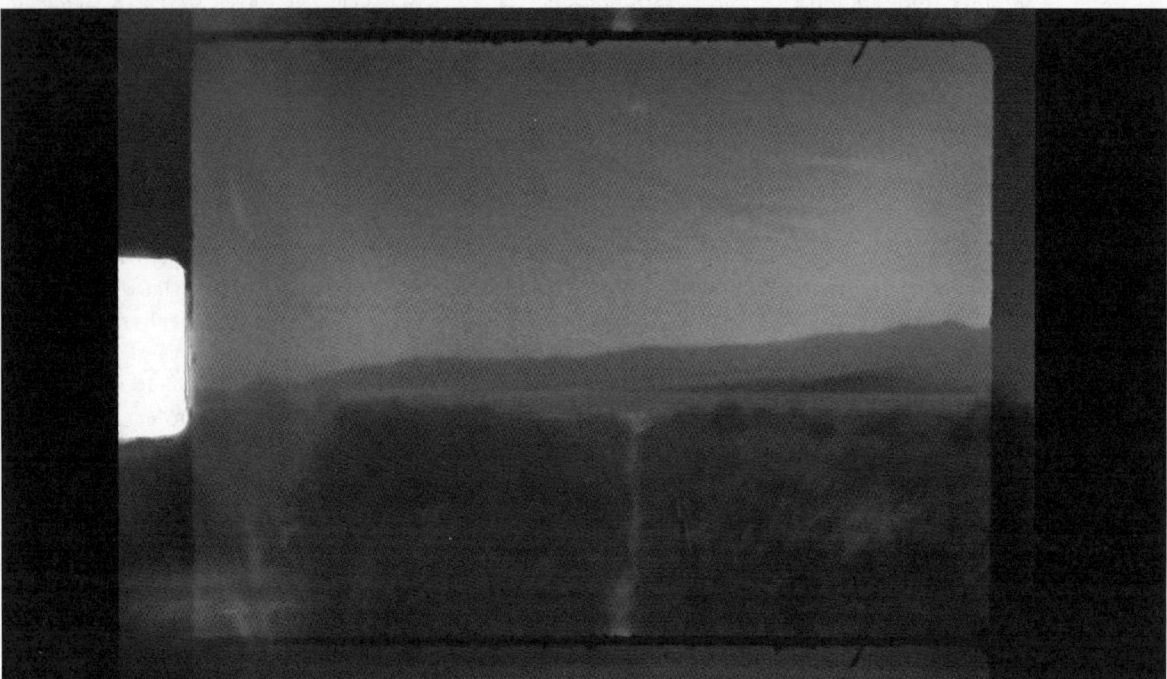

Rachel Shearer, *kapiti*, still from *Whakapapa of Shimmers*, 2024.

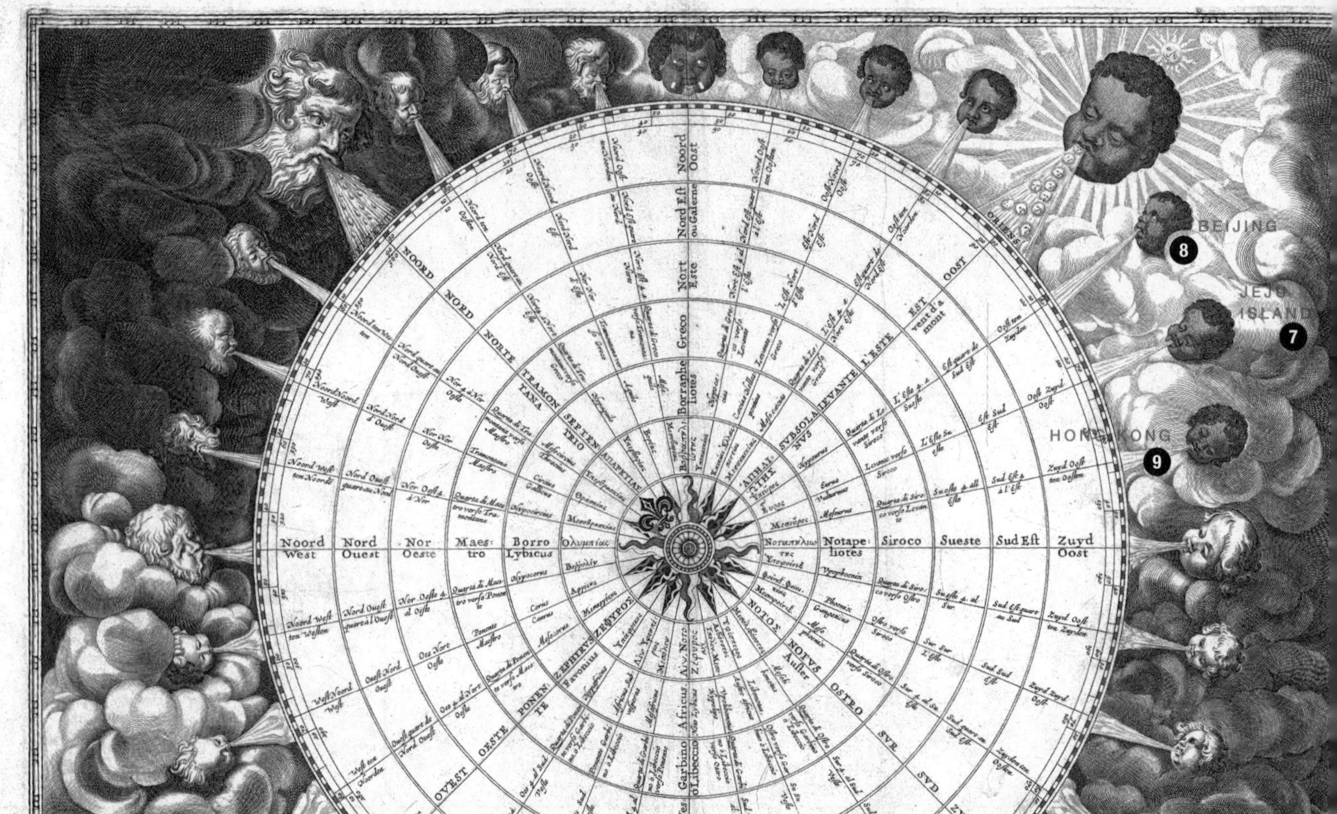

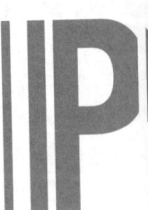

Jan Janssonius, *Tabula Anemographica seu Pyxis Nautica Ventorum Nomina Sex Linguis Repraesentans* (an anemographic chart of the winds representing the six languages), 1650.

1. Port of Los Angeles, San Pedro, California
2. Golden Gate Bridge, San Francisco
3. Seattle, Washington
4. Ōtsuchi, Iwate Prefecture, Japan
5. Hamamatsu, Shizuoka Prefecture, Japan
6. Kinshozan, Gifu Prefecture, Japan
7. Jeju Island, South Korea
8. Beijing / Los Angeles
9. Los Angeles / Hong Kong
10. Anzac Bridge, Sydney
11. Waikiki, Oahu, Hawaii
12. Orange, California

SEATTLE
3

ŌTSUCHI
4

SAN FRANCISCO
2

8 + 9

KINSHOZAN
6

LOS ANGELES
SAN PEDRO
1 **12**

5

HAMAMATSU

REPUBLIC OF WIND:

ORANGE

WAIKIKI
11

SHARED BREATH ON THE PACIFIC RIM

SYDNEY
10

by FIONA SHEN

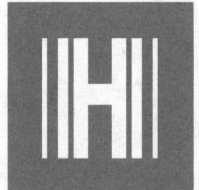ow do we perceive an expanse like the air, so often sensed as emptiness? And what if that emptiness is the air over the Pacific, the greatest expanse of ocean on earth? The constituents of air (oxygen, nitrogen, a little argon, and traces of other gases) are invisible. Air has no taste, no smell, no sound. It has heft we're usually insensible to, despite our living, as Evangelista Torricelli, inventor of the barometer, observed, "at the bottom of an ocean of air."[1] We consume it without thought, breathe 360 liters per hour of it, share it with plants and other animal species, pollute it, and disregard it.[2] But air is where we live.

We don't see or hear air under high pressure, jostling pools of low pressure to form wind. But we sense the effect of wind passing over the branches of trees, causing them to move and create vibrations in the air. We interpret these longitudinal pressure waves as sound. What we hear, then, is not air but air's effect—waves converted into electrical signals in our brains. And what we hear, not just of air but of our entire planet, is a wafer-thin slice of what is present. As biologist David George Haskell reminds us, we "live within a restricted aural world... insensitive to most of the world's vibrations and energies."[3]

If we're not to be limited by our senses, we need engineers and artists to help us grasp such elusive phenomena, to reify Nobel Prize–winning physicist Niels Bohr's pithy reminder that an unobserved phenomenon is an oxymoron.[4] We can, for instance, observe the invisible jet streams sweeping high above the Pacific through digital animations that show them braiding the globe in languorous threads of neon pink and yellow. We can trace the shifting currents of water and wind in maps that depict their motions like Van Gogh constellations. We can hear the wind in musical instruments fitted with microphones to amplify its otherwise inaudible sigh.

Artists dismantle the sensory barrier. They provide us with prosthetic ears. They magnify, sonify, and illuminate. They make the imperceptible sensate. They also remind us to relish what we can readily perceive without such interventions, creating work that allows us "to rediscover the world in which we live, yet which we are always prone to forget."[5]

1. PORT OF LOS ANGELES, CALIFORNIA: VIBRATIONS ON THE EDGE

The Port of Los Angeles is among the busiest container ports in the Western Hemisphere. Below the Vincent Thomas Bridge suspended over the port's main channel, brightly hued shipping containers are stacked like Lego bricks; massive gantry cranes slip forty-ton containers into complex configurations, muted by the roar of traffic on the four-lane bridge above. The ships are registered under the flags of mainland China, Singapore, Korea, Panama, Taiwan, and Marshall Islands. Their ports of call ricochet around the Pacific Rim: Yantian, Hong Kong, Haiphong, Cai Mep, Kaohsiung, Nagoya, Xiamen, Kobe, Ningbo, Busan. They bear names redolent of the racetrack: *Ever Forever*, *Mol Courage*, *Ever Lovely*. Their cargos might be the most pedestrian consumer goods in transit from Guangdong factories to the suburbs of Phoenix,

1. Evangelista Torricelli, quoted in Mark Vanhoenacker, *Skyfaring: A Journey with a Pilot* (New York: Penguin Random House, 2015), 136.

2. Peter Adey, *Air: Nature and Culture* (London: Reaktion Books, 2014), 7.

3. David George Haskell, *Sounds Wild and Broken: Sonic Marvels, Evolution's Creativity, and the Crisis of Sensory Extinction* (New York: Viking, 2022), 30–31.

4. Niels Bohr, quoted in John Archibald Wheeler, "Law Without Law," in *Quantum Theory and Measurement*, ed. Wheeler and Wojciech Hubert Zurek (Princeton, New Jersey: Princeton University Press, 1983), 184.

5. Maurice Merleau-Ponty, *The World of Perception*, trans. Oliver Davis (1948; London: Routledge, 2004), 39.

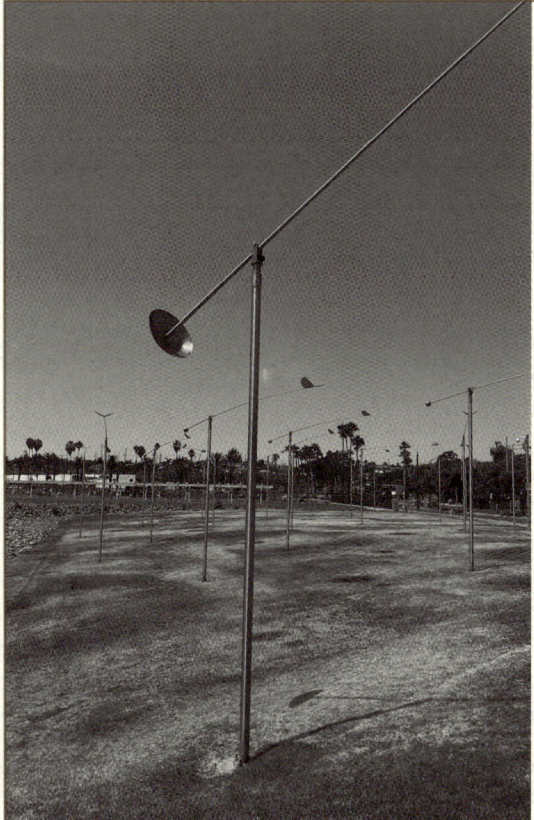

Doug Hollis, *Telltales*, 2004. Kinetic wind and sound sculptural array consisting of thirty 25-foot-long wind-activated stainless steel elements mounted on 20-foot poles, installed along the San Pedro waterfront in Southern California. Photograph by Fiona Shen.

but here, in this moment, their enterprise is achingly romantic.

A short stroll from the bridge, along the evocatively named Cruise Ship Promenade, Doug Hollis's kinetic installation *Telltales* (2004) narrates the energy of this place through motion and sound. Gigantic fishing lures bob from twenty-foot steel poles; organ pipes mounted on weathervanes moan; a metallic hum shimmers from pole-mounted wires. These are aeolian harps played by the wind, ancient instruments popular in Europe and America during the Romantic era from the late eighteenth to the mid-nineteenth centuries, rediscovered by contemporary sound artists. As air flows over the wires' surface, vortices spin from side to side causing them to vibrate. The frequency depends on their tension, diameter, and length. The wires' small surface area means that little air is displaced, but in Hollis's work, they are attached to metal bowls whose large surface areas act as resonators, vibrating in tandem with the strings and amplifying the sound waves. Hollis explains, "Both the harps and the organs operate on the same phenomena of 'sympathetic resonance,' a relationship between the diameter of the string or the volume of the tube (a piston of air) that gets excited into vibration when the wind is a certain speed. Therefore, the frequencies produced follow the harmonic series, sliding from one harmonic to another as the speed changes. The fundamental frequency (the first harmonic) is often too low or too soft to hear."[6]

The sound of the wind, like stories and epics, helps us locate ourselves in the world. Hollis's *Telltales* evokes a maritime past of tall tales and tall ships. The port's historic fishing industry finds an echo in the bobbing fishing lures that simultaneously resemble the tails of tuna or whales dipping through the waves. The organ pipes emit the low rumbles of fog horns or steamship funnels. The work creates a soundscape of the winds, ships, and people crisscrossing the Pacific moment by moment.

Historically, a port depends on wind, but here the ambient soundscape is dominated by road traffic. Hollis helps us experience a phenomenon we would otherwise struggle to hear. Such work "takes us out of ourselves and puts us into relationship with what's around us," Hollis reflects. "It takes us into the now... helps us pay attention to our existence and each other."[7] Playing aeolian music on the continent's edge, *Telltales* reminds us that wind and breath are like a shared journey to the destinations of every ship sliding past.

6. Doug Hollis, email correspondence with author, 3 August 2022.

7. Hollis, interview with author, 1 July 2022.

2. SAN FRANCISCO: BRIDGE HARMONICS

On a summer afternoon in 2020, the Golden Gate Bridge started to thrum like the nearby San Francisco Zen Center temple bells. Sometimes soothing, sometimes eerie, the bridge rang like an International Orange singing bowl, hitting a perfect A frequency of 440 Hz.[8] The hum became part of the city's auditory landscape, as much a marker of place as the steel-blue Pacific, the fingers of fog, the East Bay's sunbaked hills, and the iconic bridge straddling two peninsulas. Sound creates and is created by this place. In summer, frigid water upwells to the ocean's surface, chilling the wind that rushes inwards towards the warm interior. The wind funnels through the pinched Golden Gate Strait with a force known as the Bernoulli effect, and the bridge starts to sing.

The 1937 structure had recently been retrofitted to withstand winds up to 100 mph. Narrower slats were installed below the handrails, but the wind rushing over their sharp edges began to stimulate vortex shredding—airflow oscillating around the obstacle. The bridge became a wind instrument, amplifying the vibrations into a miles-wide soundscape.

Alerted to the Golden Gate Bridge, Highway and Transportation District's plans to dampen the vibrations with aluminum and rubber clips, musicians shared field recordings and aeolian playlists. In May 2021, Los Angeles–based guitarist Nate Mercereau and Oakland-based recording engineer Zach Parkes chose the site of the gusty Marin Headlands to record the album *Duets | Golden Gate Bridge* (How So Records, 2021), a collaboration between the bridge and Mercereau's guitar improvisations. "There's a really high note that kind of sings through everything," recalls Mercereau. "That seems to be the constant presence. For us, it was in the key of C major, and there's these lower notes that seem to swell and smear together as the volume increases or as the wind changes direction. It's not really playing with me. It doesn't know I'm there in a real, physical sense. It's like you're witnessing it happening, and I'm getting inside the sound of it."[9]

3. SEATTLE, WASHINGTON: METAPHORS OF SHARED LONGING

In Seattle, the prevailing wind blows from the southwest. The city's biggest immigrant populations come from China, Vietnam, Philippines, India, and Mexico. We can grasp this as data, but how do we experience these facts' interlocking complexity with our senses? And how might such sensory perception bring us into greater empathy with these populations? How might we be more moved by the shared fact of human movement?

The wind is an agent of migration. The wind continually shapes life and land by providing power for sailing vessels, thermals for migrating birds, and transport for seeds and pollen. Even today, an aircraft's trajectory is controlled by the wind.[10] Migrants are impelled by longing and desire but also by conflict and loss; surges of forcibly displaced peoples are a defining crisis of our century. The United Nations publishes data on the "stock" of international migrants in five-year increments. The numbers, measured according to country of

8. Erin McCormick, "The Quest to Solve the Mysterious 'Eerie' Hum of the Golden Gate Bridge," *The Guardian*, 13 June 2021, theguardian.com/us-news/2021/jun/13/golden-gate-bridge-hum-noise-san-francisco, accessed 26 October 2022.

9. Nate Mercereau, interview by Mano Sundaresan, "This Musician's Unlikely Duet Partner? The Golden Gate Bridge," *All Things Considered*, National Public Radio, 9 August 2021, npr.org/2021/08/09/1026100993/this-musicians-unlikely-duet-partner-the-golden-gate-bridge, accessed 26 October 2022.

10. Vanhoenacker, *Skyfaring*, 149–52.

11. United Nations Population Division, "International Migrant Stock 2020," un.org/development/desa/pd/content/international-migrant-stock, accessed 7 November 2022.

12. Joel Ong, interview with author and Marcus Herse, 24 August 2021.

13. Ong, *Windward | Windword*, installation with Inmi Lee at Jacob Lawrence Gallery, University of Washington, Seattle, 2016.

14. Ong, "*Windward | Windword*: Elemental Metaphors for Data Art," 25th International Symposium on Electronic Art Proceedings (2019), 564, isea-archives.org/docs/2019/ISEA2019_Proceedings.pdf, accessed 7 November 2022.

15. As Ong explains, the microphones picked up visitors' speech, did a rudimentary analysis of amplitude, and if the signal was over a floating threshold, it triggered blowers on the speakers' side of the pool. Email correspondence with author, 17 November 2022.

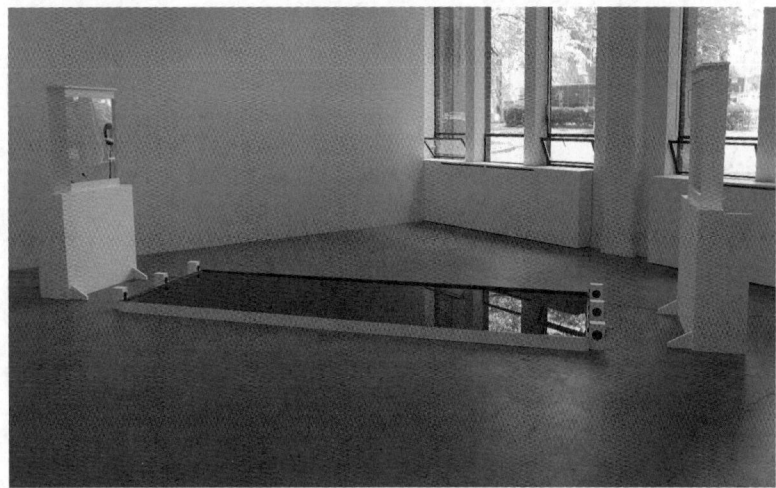

Joel Ong, *Between Us a Breeze*, 2016, installation in *Windward | Windword* at Jacob Lawrence Gallery, University of Washington, Seattle, 2016.

origin and country of destination, flow in electronic datasets, a human stock exchange that's affecting only in terms of the raw numbers.[11] But the lived experience of just one "datum" is immeasurable.

In 2015, Seattle's housing crisis coalesced with the Syrian refugee crisis. To create affective narratives based on his experiences advocating for asylum seekers and undocumented migrants, media artist Joel Ong mined data and metaphor. "I started thinking about trade winds and wind patterns, and that led to migration," Ong recalled. "How can I connect with these narratives of movement, of loss, of transience?... I started thinking of all the metaphors of freedom, of directionality, 'the wind in your sails.' The most poignant one was about the wind as a direction... as an extension of your breath."[12]

His work *Windward* (2014–ongoing) mapped metaphor to journey.[13] Conflict in one area of the world vibrates over the globe. Outside the Northwest Detention Center (NWDC) in Tacoma, where over 1,500 undocumented migrants faced deportation, Ong installed a mechanized wind vane around which 102 arrows, pointing towards the migrants' countries of origin, radiated in the sand below. At any given moment, the wind nudged the vane towards someone's homeland. An anemometer recording wind speed and pressure at the NWDC was connected to a vane installed at the University of Washington's Jacob Lawrence Gallery. A Raspberry Pi–powered screen tracked wind directions as they shifted towards detainees' homelands.

An accompanying installation, *Between Us a Breeze* (2016), was inspired by Ong's volunteer work at the NWDC and his frustration with divided "visitation booths." "I realized very early on that it was impossible to achieve a deep connection with an inmate because of a fundamental inability to 'share the air' with him."[14] In the gallery, two visitation booths encouraged visitors to speak with each other by telephone, their breath triggering ventilators that sent air rippling over a reflecting pool between them.[15]

The wind brings us together and reminds us of our connection. The air I inhale was exhaled by another. As anthropologist Tim

Ingold observes, to feel the wind is to understand that we "commingle" with everyone and everything.[16]

4. ŌTSUCHI, IWATE PREFECTURE, JAPAN: YEARNING

Wind Phone perches in a garden on a small rise overlooking the Pacific. An old-fashioned black rotary phone rests on a shelf in a glass-paned booth. The bereaved come to dial the numbers of lost loved ones and ask, "Are you well? Are you cold? Are you eating enough?" The phone cord is connected to nothing.

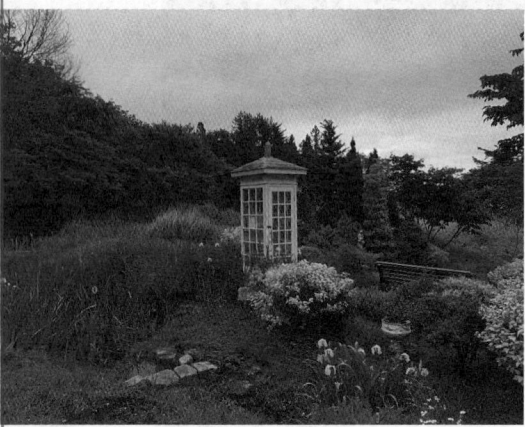

Itaru Sasaki, *Wind Phone*, 2011.

"Because my thoughts could not be relayed over a regular phone line, I wanted them to be carried on the wind, so I named it the wind telephone," says Itaru Sasaki, a garden designer who created it in 2010 after losing his cousin.[17] A year later, the nearby town of Ōtsuchi was devastated by the 2011 Tōhoku tsunami and lost a tenth of its population. The wind phone is where the survivors come to talk. Their voices journey outwards in sound waves that meet the glass, which does little to absorb them. The glass vibrates, a new source of sound. "I'll come again," ripples into the wind.

5. SENRIHAMA DUNES, HAMAMATSU, JAPAN: THE ECHO-OTAKU

Where, in time, is an echo located? Is it the moment of the original sound, or is it the reverberating after-sound? Akio Suzuki, a self-described echo-*otaku* (echo nerd) has been playing with these twists in cognition since his early twenties, when he travelled alone in the Japanese mountains, teaching himself to sing.[18] "The echo has a specific psychological effect that blurs our perception of sensing time, and our fixed notions of past and present," he explains.[19] In 1969, he invented his first Analapos (a portmanteau word blending analog and postmodern) from two large juice cans and a metal spring.[20] It was a rudimentary instrument reminiscent of the tin-can telephones children create, but it framed the concept of play, or *asobu* (a word he often uses), as a path to sharing the non-human sounds in our environment that we're otherwise unattuned to.

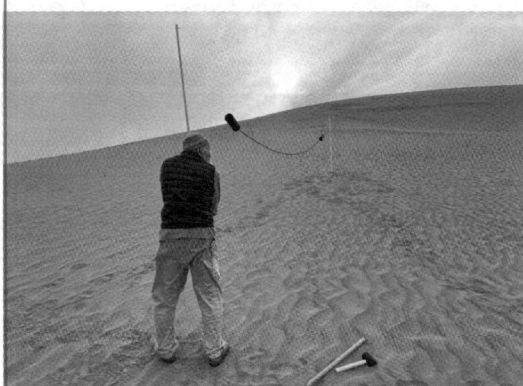

Akio Suzuki playing the Analapos at the Tottori Sand Dunes, Tottori, Japan, March 2023. Photograph by Lawrence English.

16. Tim Ingold, "Earth, Sky, Wind, and Weather," *The Journal of the Royal Anthropological Institute* 13 (2007): S29.

17. Miki Meek, "Really Long Distance," in "One Last Thing Before I Go," *This American Life*, 24 April 2020, thisamericanlife.org/597/one-last-thing-before-i-go-2016/act-one, accessed 10 November 2022.

18. "Tomoko Sauvage, "Magical Stones: An Interview with Akio Suzuki," *The Wire*, April 2019, thewire.co.uk/in-writing/book-extracts/magical-stones-interview-with-akio-suzuki-by-tomoko-sauvage, accessed 18 October 2023.

19. Akio Suzuki, interview with Elisa Ferrari, "Sound as a Space—An Interview with Aki Onda and Akio Suzuki," trans. Emma Metcalfe and Ruimin Li, *Western Front*, 18 October 2017, legacywebsite.front.bc.ca/events/sound-as-a-space-an-interview-with-aki-onda-and-akio-suzuki/, accessed 18 October 2023.

20. Sauvage, "Magical Stones."

21. For a detailed and lyrical exploration of this, see Haskell, *Sounds Wild and Broken*.

22. Eisuke Yanagisawa, interview with author, 18 March 2022. Subsequent quotes are from this interview.

Over the next decade, Suzuki experimented with variations of the instrument, becoming ever more sensitive to the sculpting of sound by unique environments. His 1982 performance *Analapos* on the Senrihama Dunes in Hamamatsu erases notions of time and physical place like wind erases dunes. The hills he trudged up with his Analapos no longer exist, nor do the ripples of sand over which the twin canisters hovered, whistling in the wind. It is the nature of mountains to erode and be carried down rivers to the sea. It is the nature of dunes to dissolve. In the video footage of this performance, this *asobu*-lover is laughing at the paradox of the music he hears in the echo of erasure.

6. KINSHOZAN, GIFU PREFECTURE, JAPAN: HEARING NON-HUMAN HISTORIES

With different listening organs, we might hear the world as a cacophonous web of information: insects blasting vibrations through their legs onto plants, shrimp snapping underwater, fish vibrating their swim bladders to squeak, knock, grunt, moths rubbing their papery wings together, the ultrasonic clicks of bats, the infrasonic rumbles of elephants. If we heard it all, the din would fell us.

All species have their sensory boundaries, drawn through evolution to specific ranges of frequencies.[21] Ours is relatively narrow, ranging between two and four kilohertz. To nudge that range a little wider, artists like the Kyoto-based ethnographer and filmmaker Eisuke Yanagisawa use instruments like hydrophones, bat recorders, and microphones to open our ears and minds to "a less anthropocentric way of perceiving the world."[22]

It's in our nature to personify, to attribute human behaviors and desires to other species and even natural phenomena. In Greek mythology, the force of the wind was personified as Aeolus, who corralled the winds on the island of Aeolia and controlled their paths. Yanagisawa upturns this, restoring the wind's nonhuman agency in his *Path of the Wind* (Gruenrekorder, 2018), an album of seven field recordings of aeolian harps in often remote locations in Japan.

Wind moves as a fluid, flowing around and over obstacles, swirling, eddying. The unique topographies of place change its course and speed, making these recordings documents of a distinct place in time. Yanagisawa explains that his simple handmade harp, fitted with two microphones, "is an instrument, but it's also a microphone to sonify inaudible sounds... The box functions as a resonator so we can hear the sounds at amplitude... Different places resonate differently with the harp," meaning that we are given access not just to a place in the few minutes

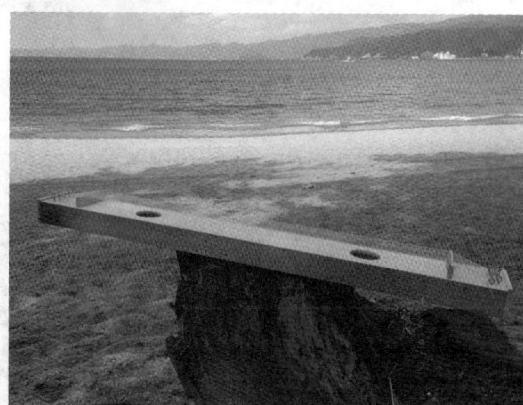

Eisuke Yanagisawa's handmade harp captures the sound of the wind for "Seagull," from *Path of the Wind*, 2018. Recorded on Kehi no Matsubara, Tsuruga Peninsula, Japan. Photograph courtesy the artist.

of the recording, but to the very path of the wind over time.

"Every place has a special history, so I try to imagine, I try to let the listener imagine the timescale of the place. For every place, the sound always has a connection with the past." Sometimes that past is recent, such as in "Seagull," recorded under the Tsuruga Peninsula near several nuclear power plants. Sometimes the timescale is outside ours, such as in "Old Camellia Tree," recorded beside a 1,200-year-old *Camellia japonica*: "The tree is part of the environment. The wind is made by that environment. The tree is part of making the wind." The harp's mellow bronze tones, never shifting dramatically from mid frequencies, suggest not a tree in the moment but its continuous 1,200 years of presence.

Other works capture geologic time. "Kinshozan" was recorded on a small mountain in Gifu Prefecture, famous for its Triassic-era fossils of corals, bivalves, and snails, where limestone has been mined since the Edo period (1603–1868 C.E.). Yanagisawa placed his harp at a watchtower overlooking the mine and recorded the complex textures of this multi-strata place, layering the high-frequency whistles of the harp strings over the insistent mechanical tapping of mine drills. Underneath is a rumbling bass like an underwater sigh, a specter, perhaps, of earth's early ocean animals sensing vibrations in the water. Despite our sensory handicaps, our ears have been opened to other ways of being.

7. JEJU ISLAND, SOUTH KOREA: WIND LIKE A REQUIEM

For five months in winter, Jeju Island, off the southern tip of the Korean peninsula, is scoured by northwest winds from Siberia. Parts of Jeju are so windy year-round that the island hosts South Korea's first offshore wind farm and showcases the nation's clean energy ambitions.[23] While the massive turbines capture energy to generate electricity, Tai-

23. Byeongtaek Kim, et al., "Offshore Wind Resource Assessment Off the Coast of Daejeong, Jeju Island Using 30-Year Wind Estimates," *Scientific Reports* 12, no. 14179 (2017), nature.com/articles/s41598-022-18447-7, accessed 19 August 2022.

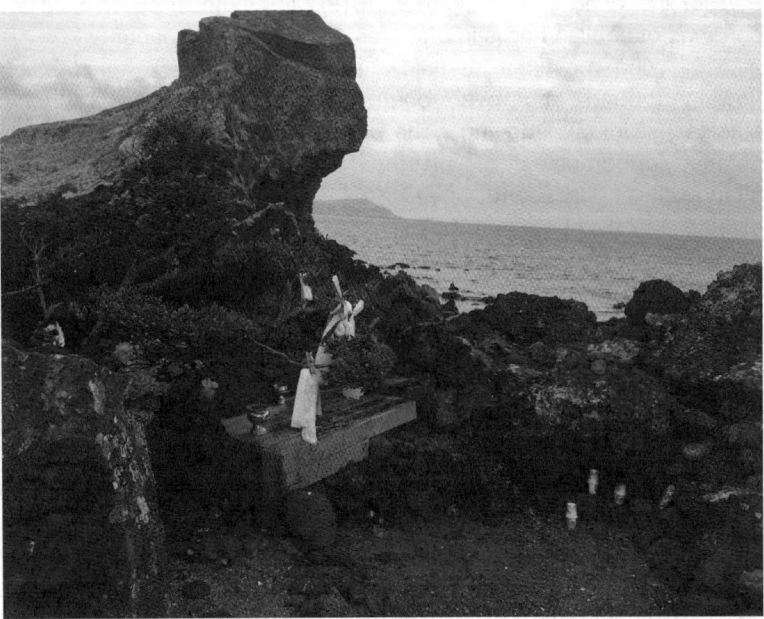

Set up of Hong-Kai Wang's field recording for *Borom*, Jeju Island, South Korea, 2020. Photograph by Kang Kyung Duck, courtesy the artist.

24. Kim Sijong, interview with Hong-Kai Wang, 13 March 2020, in "Booklet/ Playbill for Borom," The National Culture and Arts Foundation in Taiwan and Jeju Biennale (2020), 16.

25. Ibid., 17.

26. Ibid.

27. Hong Kai-Wang, interview with author, 30 August 2021.

wanese artist Hong-Kai Wang harnessed Jeju's winds to generate empathy.

Her 2020 work *Borom* ("wind" in the Jeju dialect) uses field recordings to memorialize the flight of refugee poet Kim Sijong to Japan in 1949 following the Jeju Uprising. This popular revolt turned the island into an abattoir. An estimated 30,000 islanders (ten percent of the population) were massacred during the brutal suppression of the pro-democracy movement by the South Korean dictatorship aided by the U. S. military. Sound recordings trace not just the geography of Kim's historic journey but the inner landscape of his fear, despair, horror, and hope.

A volcanic island, Jeju's soundscape is shaped by wind, water, and basalt rock. Beyond its status now as a tourist destination, its identity is still wrapped in its windscape. "The sound of its wind wraps and rings in your ears," insists Kim, "and even on days that are clear and winds are calm, you can hear deep vibrating sounds in the air.... And there are always winds passing through the cracks of the stone walls. Since the stone walls consist of volcanic rocks and contain king-sized and small holes, many complicated sounds emerge together from the winds like a symphony."[24]

Seventy-two years after the massacre, it is this soundscape through which Kim recalls events "too painful to describe in words."[25] At Sarabong, captured villagers were dumped at the foot of a hill, transported out to sea, and thrown in the ocean, wired together by their wrists. "The gravel beach was rumbling," recalled Kim. "Even today, when I think of the Jeju Uprising, I vividly hear the sounds of waves beating against the gravel beach and rocks, making a go-go sound.... I remember hearing the sounds of waves breaking on the beach, the gravel beating together, and the wired corpses being scraped in and out against the gravel beach."[26] *Borom*'s recording on this location is a layering of the low-frequency rumble of the wind, above which waves suck and crunch the gravel shore. Above this, a bird asserts its song as if in witness. It is ominous and insistent. "Sound can be felt," argues Wang. "Sound doesn't stand alone. What matters more is the perceptual space the sound evokes."[27]

In early June 1949, Kim escaped on a small boat to Guantal Island, a low volcanic reef about seventeen miles offshore. For four days, he huddled in a crack in the basalt. Again, over seventy years later, the recollection of the wind was the key to memory: "Strong cold winds rushed in from the north. The hyoong, kyoong sounds would break through the crack and sprayed swirls of the sea into where I was hiding; it is difficult to describe how persistently the wind blew through the night... I was still wet from winds blowing and waves splashing against the crack. I was freezing. The depth of fear and darkness I was feeling is hard to explain. It was horrifying. Am I a bug? No, I wasn't a bug. I was less than one. I wondered, why was I breathing?" The field recording from Guantal weaves together the same sound sources as the one from Sarabong, but here they are subtly different—the wind murmurs, the waves roll off basalt, and a gull cries like a child. Kim recalled, "The wind was not a sound that flowed straight in one direction but a sound that is blunted with every swell of a wave."[28] Kim's eventual rescue, his asylum in Japan, and his continued breath are intimated in the hushed respiration of this recording.

Borom posits that the past is still present. Memories of the Jeju Uprising are still embodied in the wind whistling across porous granite. Specters of the manacled dead manifest in the sound of waves raking the gravel. "Sounds are like ghosts," the artist-philosopher Salomé Voegelin reminds us. "They slink around the visual object, moving in on it from all directions, forming its contours and content in a formless breeze."[29] For a long time, their presence was ignored. Jeju became a resort island, stippled green with golf courses and luxury hotels. But trauma reverberates through generations and is as difficult to ignore as the wind that cuts short conversations. The island is gradually coming to terms with its past; in 2006, the South Korean government apologized for the butchery and opened a park honoring the victims two years later. *Borom* takes its place in this process. "This is about trauma and healing," explains Wang. "I wanted to open another way to access another's pain."[30]

For Kim, the wind's sound "pierced through my bones."[31] Decades later, the sound waves from *Borom*'s recordings of gravel beaches and basalt hideaways enter our bodies and transform us.

8. LOS ANGELES / BEIJING: THE ENERGY OF BREATH

The Chinese alchemist Tao Hongjing (456–536 C.E.) was an adept wind listener. Just by attending to the wind in a grove of pine trees, he could "spontaneously reach an immortal's state."[32] But as the ancient philosopher Zhuangzi explains, this type of listening is achieved with the ears and mind and the spirit *qi*.[33] *Qi* is more a verb than a noun—a transforming energy that informs "an organic, holistic, and enchanted worldview that the cosmos and the myriad things (including humans) are a correlated organism that are constantly resonating, condensing, disintegrating, and forming unity."[34]

Breathing, we manifest *qi*. Blowing, the wind manifests *qi*. In the pre-Qin period (2100–221 B.C.E.), the wind was *qi*.[35] Its simplified written Chinese character resembles breath as a vibrating exhalation: 气.

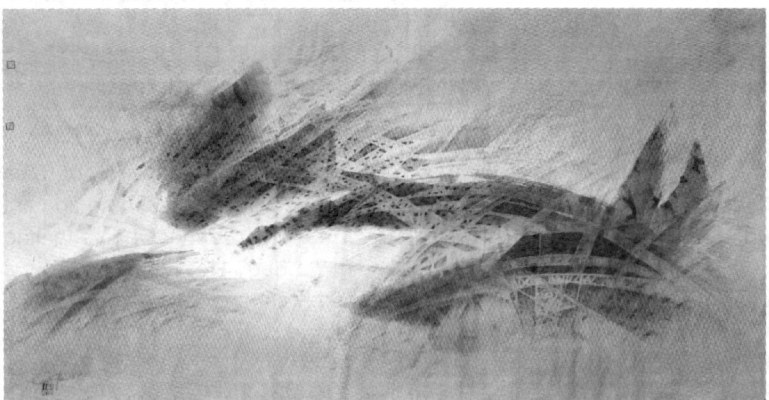

Cai Xiaosong, *#5*, 2023. Ink on paper, 70 7/8 x 35 7/16 in.
Photograph courtesy the artist.

28. Kim, interview with Hong-Kai Wang in "Booklet/Playbill for *Borom*," 19.

29. Salomé Voegelin, *Listening to Noise and Silence: Towards a Philosophy of Sound Art* (New York: Continuum, 2010), 12.

30. Hong-Kai Wang, interview with author, 30 August 2021.

31. Ibid.

32. Excerpt from a poem by Lu You, quoted in Susan E. Nelson, "Picturing Listening: The Sight of Sound in Chinese Painting," *Archives of Asian Art* 51 (1998/1999): 37.

33. Nelson, "Picturing Listening," 33.

34. Jing Wang, *Half Sound, Half Philosophy: Aesthetics, Politics, and History of China's Sound Art* (New York: Bloomsbury, 2021), 21.

35. Ibid., 36.

36. Cai Xiaosong, interview with author, 24 April 2021.

37. Bovey Lee, interview with author, 6 April 2022.

Trained in traditional Chinese ink painting and calligraphy, Cai Xiaosong listens to nature as he paints. In his Beijing studio, he listens to singing insects. In Los Angeles, he listens to the ocean and recordings of birdsong and wind in forests—not to reach an immortal's state, but to help attune himself to nature's energy: "to be in awe."[36]

Earlier in his career, informed by the traditional Chinese *shanshui* (landscape) tradition, his work leaned towards the representation of rocks, mountains, and planets. Now, as an extension of that tradition, he paints the *qi* energy pulsing through them. Sweeps of layered ink reverberate and drift into emptiness, embodying cycles of incipience and transformation. "Energy comes from the collision of opposites," he remarks. Wet ink on dry paper; the philosophical idea with the physical act of creation. "When you bring things together, they create their own energy," Cai continues. In the historical treatise the *Book of Han* (c. 82 C.E.), wind is explained as just such a collision of opposites—the *qi* from earth thrusting against the *qi* from heaven—a metaphysical explanation for the meteorological description of solar-heated air rising, cool air sinking.

Painting and calligraphy are means of self-cultivation. Like breathing, they are a way to manifest *qi*. Beyond this, by depicting the vibrating energies of life, Cai's work evokes our own awareness of the same energies. It reminds us we share resonant being and that maybe, after all, we can realize our own immortality.

9. LOS ANGELES / HONG KONG: CONNECTION IN ISOLATION

So much of Bovey Lee's imagery is airborne: paper planes, balloons, streamers, birds, sailboats, and migrating insects dependent on the wind. In a recent work *To Return* (回) (2021), wind-driven waves crash towards a portal created by the Mandarin character 回. While this intricate papercut was executed in Los Angeles during the pandemic, in Lee's imagination, she was gazing over the Pacific towards her Hong Kong birthplace. "As an immigrant, I relate my experience with images that are in flight, that are suspended, that are floating, or constantly moving."[37] The wind is symbolic of adventure and quest (an association that inhabitants of the small, densely populated Kowloon Peninsula and islands readily appreciate). But it's also a metaphor for the breath that flows through her work. "When I cut paper, why I am so drawn to it, even though it's so laborious, is because it connects me to my body and my breath as a living being. I feel so alive when I'm creating cut paper… that sense of presence is so important, and the *qi* is what can bring you there." *Qi* flows through body flows through art flows across the Pacific. Perceiving *qi* returns us to ourselves.

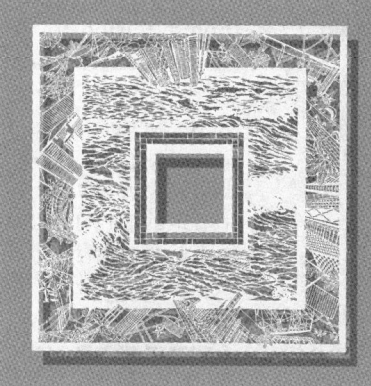

Bovey Lee, *To Return* (回), 2021. Chinese Xuan paper on silk, 33 x 33 in. Photograph courtesy the artist.

10. **SYDNEY:**
ANIMATING THE INANIMATE

As an art student during the mid-1990s, Jodi Rose speculated that Sydney's new Anzac Bridge looked like a giant aeolian harp. Inspired by Joyce Hinterding's encouragement to listen to sounds in space, she sought the help of the Australian Broadcasting Corporation (ABC) to record its cables playing in the wind before it was opened to traffic. The 1995 recording, made by placing small contact microphones directly on the cables, is now an important archival record; for Rose, it initiated a global journey to record the world's "singing bridges." To date, she has a database of over eighty.

The initial recording of the Anzac Bridge is a voluptuous sound sculpture. Its foundation is a low-frequency rumble, in which, Rose notes, we "hear the materiality of the bridge in space."[38] Above, cables under tension zing and pop like rain hammering on a metal drum. And floating on top, resonances like bells. "It's quite a profound moment to stop in the middle of the bridge and listen to the sound. Everything fades away, and it's a very intimate experience because even though it's a big, noisy, urban environment, the sounds can be delicate. It's genuinely transformative." In 2004, as an ABC artist-in-residence, she created more elaborate compositions from her initial recordings with composer Ben Fink, layering extended improvisations by cellist Ion Pearce and saxophonist Trevor Brown. These "Songs of the Bridge" are intricately dimensional in their evocation of multispecies braided sonics.

In Rose's worldview, a bridge is the opposite of inanimate. An extension of these lightly edited field recordings has been the composition of a global bridge symphony, an ongoing site-specific project in which the world's bridges play together. "I imagine the cables having a secret language so that the vibrations inside the structure are something that's not necessarily translatable by human ears, but something the bridges can hear and understand." It's a profound "otherness" to which Rose wishes to draw our attention. "I feel there's a need for a sense of connection with the world beyond the physical."

Bridges have long been considered a pinnacle of human engineering, and activating their sonic qualities depends on the mediation of technology to translate vibrations into sound within our audible range. But bridges also vibrate at infrasonic frequencies too low for us to discern, and "they sing whether we're listening or not." Far from enshrining our status, these structures put us in our place. Jane Bennett describes such encounters as "moments of sensuous enchantment with the everyday world—with nature, but also with commodities and other cultural products."[39] The music of found aeolian sculptures such as bridges is often described as extraterrestrial, but attentive listening restores us to the magic of the worldly.

11. **WAIKIKI, OAHU, HAWAII:**
A ROOSTER'S BREATH

It is difficult to retrieve the mythical heart of Waikiki. The high-rises, multinational hotel chains, and tourist restaurants are a distraction. There is plenty of fried, roasted, curried, kabobbed chicken. But little

38. Jodi Rose, interview with author, 16 March 2022. Subsequent quotes by Rose are from this interview.

39. Jane Bennett, *Vibrant Matter: A Political Ecology of Things* (Durham, North Carolina: Duke University Press, 2010), xi.

40. Jhumpa Lahiri, *Whereabouts* (New York: Alfred A. Knopf, 2021), 93.

41. For a discussion of rhythmical time created by sonic environments, see Eleni Ikoniado, *The Rhythmic Event: Art, Media, and the Sonic* (Cambridge, Massachusetts: The MIT Press, 2014).

42. Tsuyoshi Hisakado, interview, "On Hawai'i Triennial 2022," Ota Arts, https://www.otafinearts.com/viewing-room/40-interview-tsuyoshi-hisakado-on-hawaii-triennial-2022/, accessed 30 November 2022.

43. Alan Marsden and Richard Leadbetter, "Music: Seeing and Feeling with the Ears," in *Sensory Arts and Design*, ed. Ian Heywood (London: Routledge, 2017), 157–71. See also Haskell, *Sounds Wild and Broken*, for the concept of sound as generative.

that tells how the magical rooster Ka'auhelemoa flew from the nearby Palolo Valley to Waikiki Beach and scratched in the sand. The chief of Oahu recognized it as an omen and planted a grove of coconut trees. It became the site of a royal residence, sonified by the wind whispering through 10,000 coconut trees.

Kyoto-based artist Tsuyoshi Hisakado distills the historical residue from sites like this that we think we know or overlook as mundane to conjure the ambience of places still tethered to their past. His work *Pause*, installed in Waikiki for the 2022 Hawaii Triennial, parses place through light, sound, and motion. An enclosed, airy room cycles between darkness and ambient blue light referencing the Honolulu sea and sky. Spotlights momentarily illuminate white georgette curtains billowing in the breeze from hidden fans; the lights click on and off with the rhythm of surf, accompanied by a low rumble of sound that amplifies to a crescendo like a breaking wave. In the blue light, the spotlights form discs of sunset copper. In this minimalist work, the surrounding cityscape is scraped away, and the past of supernatural roosters and wind playing through expansive coconut groves converges with the present space of listening. The sound is abstract; we can't identify its origin, but in this blue cocoon, it evokes the wind and surf from which we are well-sheltered, an affect poetically described in Jhumpa Lahiri's novel *Whereabouts*: "Outside, there's a ferocious noise coming from the crashing of the waves and the roar of the wind: a perpetual agitation, a thundering boom that devours everything. I wonder why we find it so reassuring."[40]

Our embodied listening prompts us to rethink our connection to place and the relationship between time and space. We find ourselves in rhythmic rather than chronological time.[41] Screened from external distractions, we are in a recurring present moment that is a "pause" from the frenzied touristic experience of contemporary Waikiki. Poignantly, as the artist explains, it marks a pandemic-induced pause from the life we expected to live.[42] A pause is a global burden, but Hisakado shows us it is also an opportunity for a heightened multisensory experience where sight and sound are not separate. Hearing is not just a sonic experience as our senses recruit each other in acts of creativity.[43] From Hisakado's sparse elements, we imagine a world into existence.

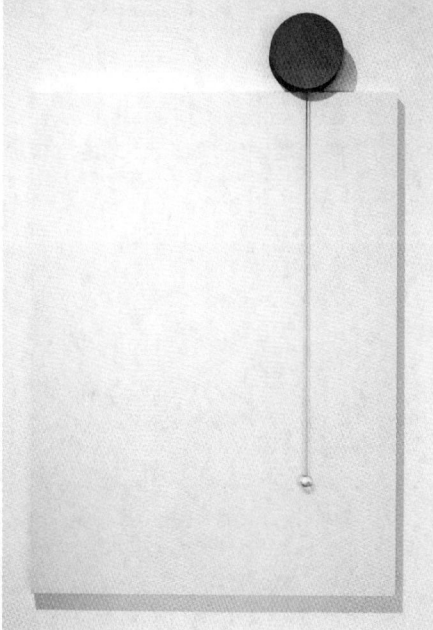

Tsuyoshi Hisakado, *Pause*, 2022. Brass, stainless steel, lens, clock movement, battery, paper, and panel, 36 11/32 x 23 27/32 x 1 5/8 in. Photograph © the artist, courtesy Ota Fine Arts.

12. ORANGE, CALIFORNIA: WAITING ON THE WIND

In Southern California, we have little trouble perceiving the malevolence of the Santa Anas—the hot, dry katabatic winds that rush down to the coast from high elevations, howling through mountain passes, plaguing our fall seasons, and knocking out the power. Joan Didion loathed them as markers of catastrophe and apocalypse, "blowing up sandstorms out along Route 66, drying the hills and the nerves to the flash point."[44] "Every booze party ends in a fight," observed Raymond Chandler of these notorious fire-spreaders and inciters of mayhem, murder, and migraines.[45]

In Southern California, we are convinced we are on intimate terms with this "devil wind." Visual artist Virginia Katz convinces us they are celestial. Between 2000 and 2008, she created a series of collaborations with the wind for which she tied fine-tipped rapidograph pens to strings attached to a *Melaleuca quinquenervia* (broad-leaved paperbark) tree in her Orange County garden and, over periods of several hours, allowed them to draw portraits of different wind conditions onto paper weighted to the ground. The onshore and offshore winds, drawn in metallic ink on black paper, look like constellations, their subtle webs of gold and copper punctuated by dense pulses of light.

Like aeolian sound sculpture, these visual works distill wind's path. They trace the meteorological conditions underlying the Santa Anas as the winds drop from desert to coast, reaching the suburban landscape around Katz's home and the tree in her garden (the species chosen for its fine, flexible branches), activating the knocking, jostling strings and pens. If not celestial, the origins of these winds are high in the deserts of the Mojave and Great Basin over Nevada and Utah. The difference between this high-pressure mass of cold air and the solar-warmed low-pressure air mass over the Pacific causes the desert air to sink, becoming warmer and drier as it gathers downward momentum. Reaching the mountains that divide the desert from the coastal regions, they funnel through passes and canyons, becoming hotter, drier, and dustier.

Satellite imagery captures the dramatic swathes of dust over the coast. Color-coded animations of these multiday events depict the region pulsing between red and blue, wind-speed arrows churning.[46] Katz captures them

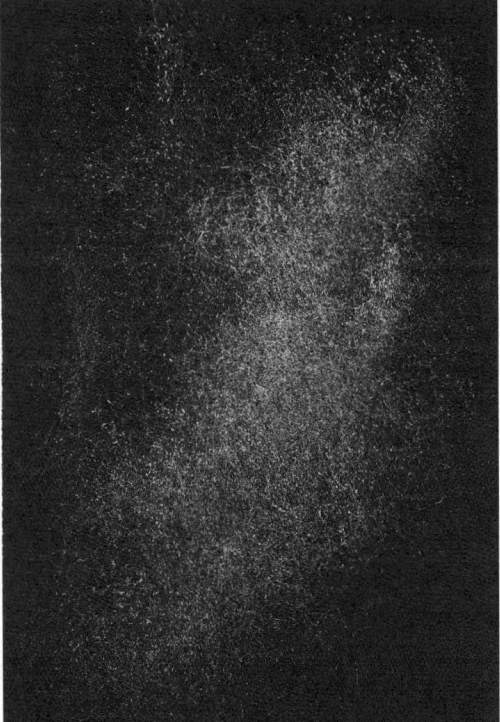

Virginia Katz, *WIND, On-Shore Flow, 7 Hours of Observation, Green and Blue, 3/28/08*, 2008.

44. Joan Didion, "Los Angeles Notebook," in *Slouching Towards Bethlehem* (1961; New York: Farrar, Straus and Giroux, 1990), 217.

45. Raymond Chandler, quoted in Didion, "Los Angeles Notebook," 218.

46. See, for example, those created by Robert Fovell, University of California, Los Angeles: people. atmos.ucla.edu/fovell/ ASother/mm5/Santa Ana/winds.html, accessed 24 November 2022.

47. Virginia Katz and Eric T. Vogler, "Wind Art Physics," n.d.

48. Katz, interview with author, 31 March 2022.

49. Katz, conversation with author, 29 November 2022.

as a delicate tracery of longitudinal lines, oscillations, stutters, feathers, skips, and rotations embodying the wind's journey. Her collaborator, environmental physicist Eric T. Vogler, extrapolated the equations behind some of the marks created by the wind's direct action on the strings or indirectly by the quivering branches.[47]

"There's a language of mathematics and physics in these drawings," remarks Katz.[48] The Santa Anas have been implicated in some of Southern California's worst recent wildfires, but so have human-induced climate change, land development, and malfunctioning utility equipment. "My work seeks to heighten our awareness that we're intricately involved in the space around us," she explains. "We'll see ourselves and how dependent we are."

Katz drew on her endurance as a marathon runner to monitor her wind drawings for stretches of up to ten hours. "When you spend many hours waiting, something magical happens. The whole world starts opening up. I started noticing the bees pollinating the tree's white bottlebrush blossoms. I saw the pollen dropping onto the drawing. There are other ways of translating what the landscape can be besides what we take for granted."[49]

Her wind drawings offer a different engagement with the world that helps us transcend our human-centric ideas of immanence or absence, beauty or ugliness, utility or worthlessness. "I'm allowing nature to be itself," she concludes.

Itself, in all its shimmering, vibratory presence. <

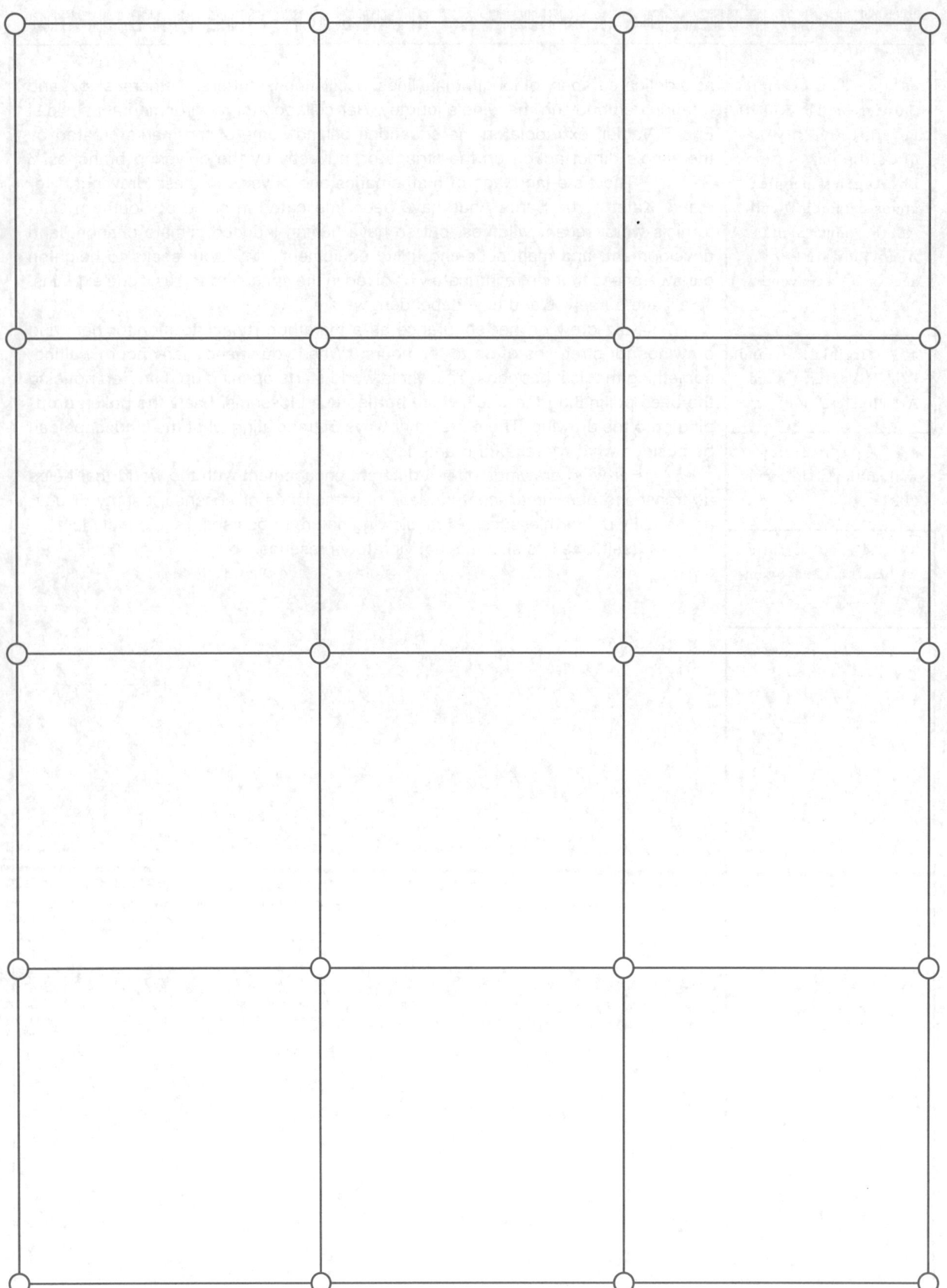

LAUREN BON AND

Lauren Bon, an environmental artist based in Los Angeles, is the driving force behind the Metabolic Studio. Her studio focuses on developing self-sustaining and diversifying systems of exchange that rejuvenate the ecological web, including a unique sonic practice.

Since *Not a Cornfield* in 2005, Bon has been involved in transmission art, notably with locations along the Los Angeles River's cyborg watershed. This sonic network spans from Owens Dry Lake to the Salton Sea, areas both known for their ecological challenges. Her work highlights the role of vibration in activating living systems, exploring how sonic frequencies can impact ecological health.

In 2023, Bon utilized construction for her *Bending the River* infrastructural intervention to create *The Great Vibration*, an exploration of how vibration can transform stagnant energies. This piece is an analog field recording captured during the installation of a well adjacent to the Metabolic Studio. The recording documents the process as sixty-foot shoring panels were vibrated into the ground, capturing the intense energies resonating through the studio.

Directed by Bon and realized in collaboration with field recordist Ian Wellman and filmmaker Maurício Chades, *The Great Vibration* presents an immersive soundscape that transitions from foreboding and violent to meditative and ecstatic. <

THE METABOLIC STUDIO

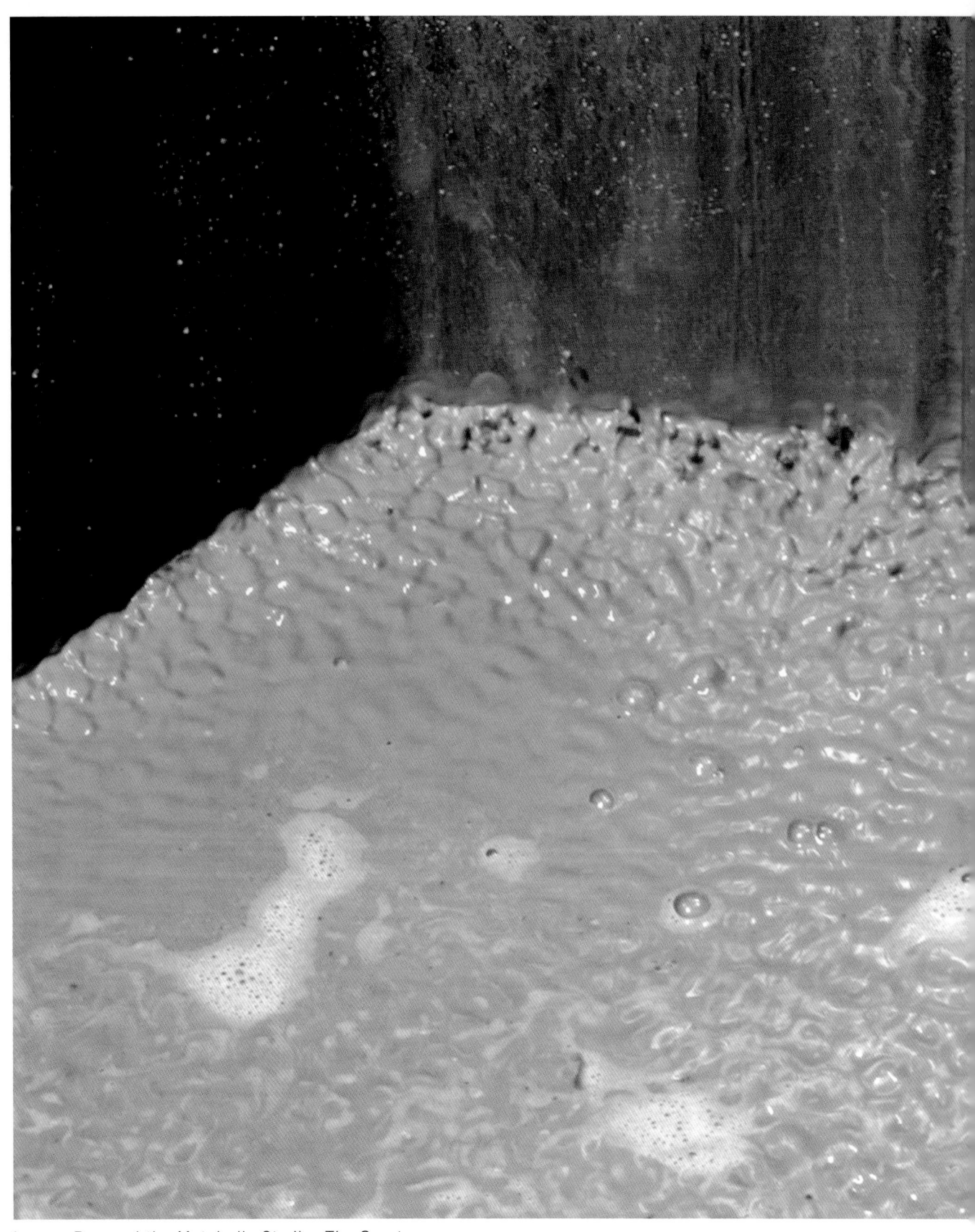

Lauren Bon and the Metabolic Studio, *The Great Vibration*, 2023.

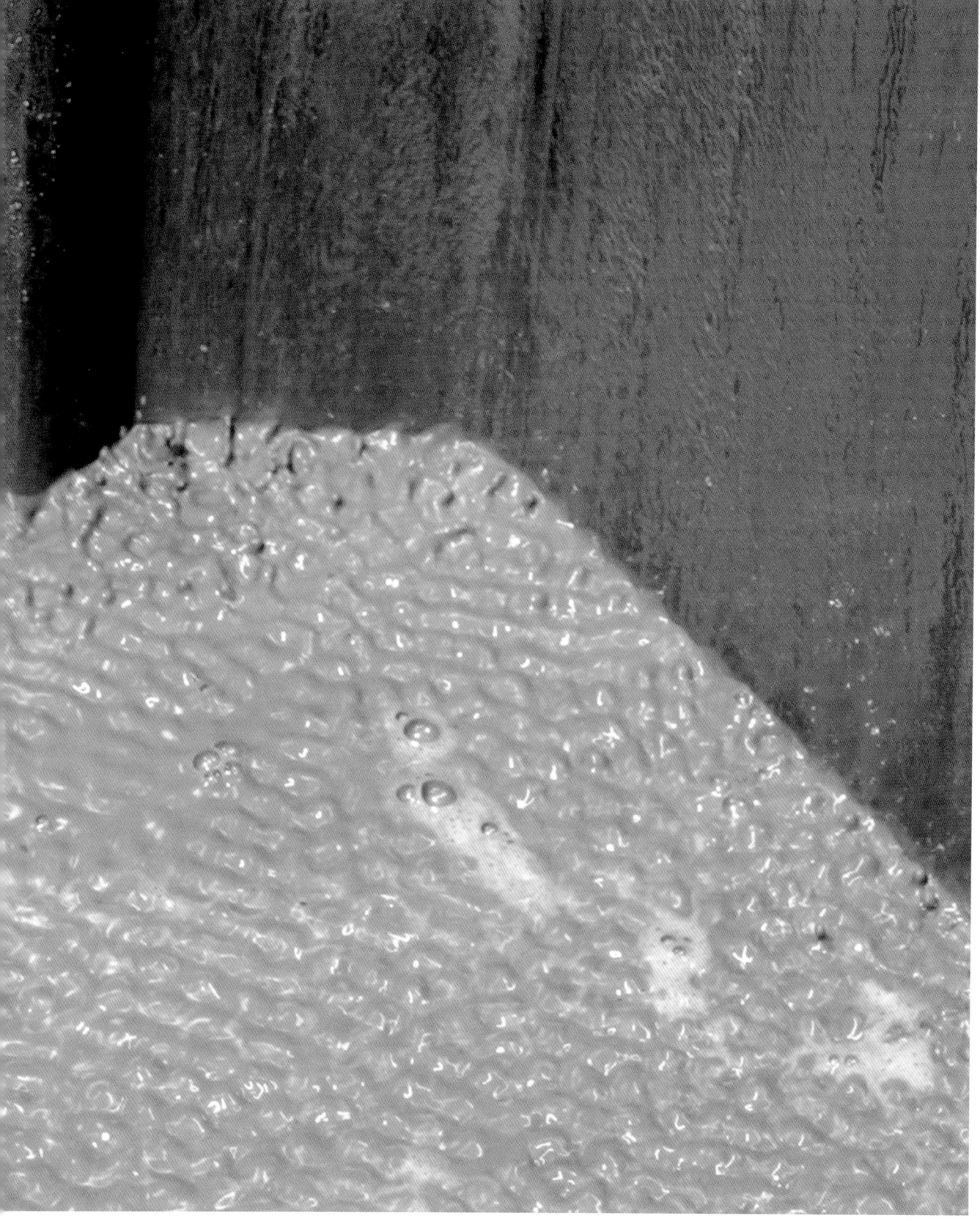

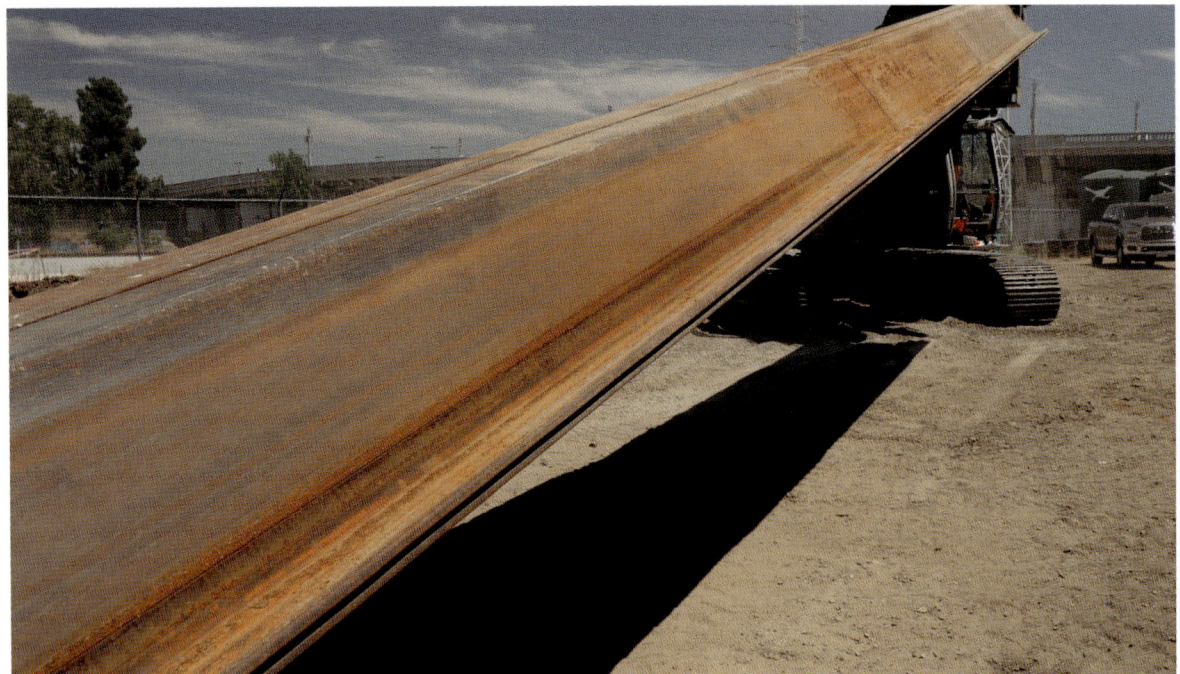

Lauren Bon and the Metabolic Studio,
60-foot-long steel shoring panel about to
be inserted into the ground for *Bending
The River*, 2023.

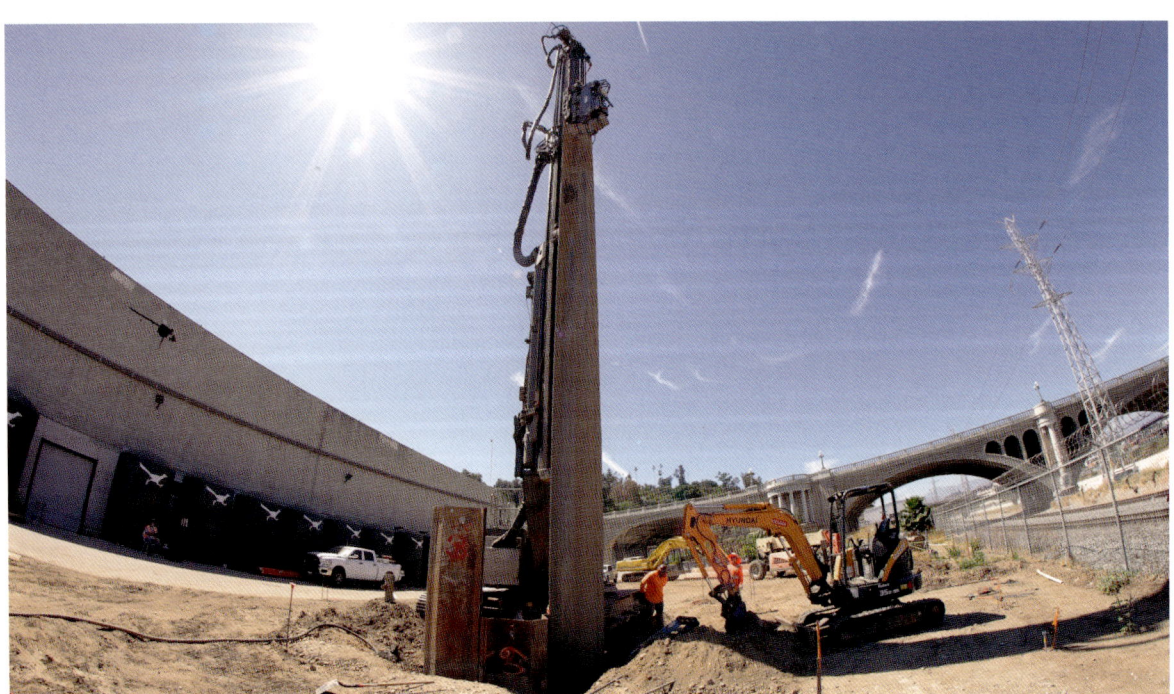

Lauren Bon and the Metabolic Studio, shoring panel being inserted into the ground
using an ABI vibratory drilling machine for
Bending The River, 2023.

JENEEN FREI NJOOTLI

Working in their home territory of Old Crow, Yukon, Jeneen Frei Njootli is a two-spirit/queer artist whose work engages sculpture, performance, music, textile design, and feral scholarship. Through public sound and performance works, they create intimate and embodied experiences related to ancestry, cultural heritage, and the land and its human and nonhuman inhabitants.

For *Energy Fields: Vibrations of the Pacific*, Frei Njootli created a performance work rooted in their research on the unceded Indigenous territories bounded by the provinces of western Canada. For *Bone meets blade, sonified calling out. who what will utter back* (2023), they "sonified" the process of making a traditional hunting instrument out of a caribou shoulder bone. The territory's Indigenous people use these instruments to make sounds that attract caribou herds during hunts. Using a Dremel grinder and wearing a single-beam headlamp, Frei Njootli modified the bone's central ridge while contact microphones, distortion pedals, and an electric guitar amplifier magnified the sounds of their craftsmanship—that is, the instrument's becoming. The performance concluded with Frei Njootli playing the altered shoulder bone; they sang into its new form, their voice activating the resonant character of the bone and producing an instance of the titular calling out. <

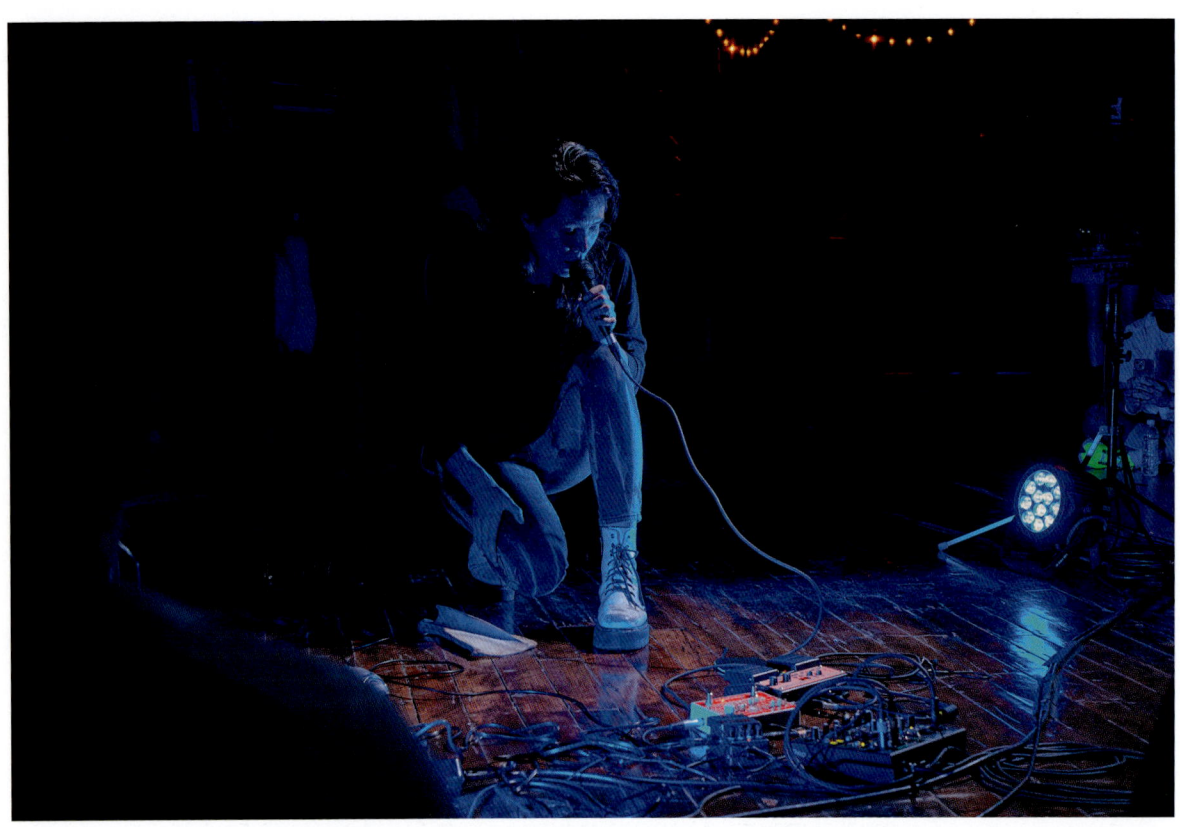

Jeneen Frei Njootli, *Bone meets blade, sonified calling out. who what will utter back*, 2023. Performance at Production Club, Los Angeles, 2023.
Photographs by Gabriel Sweet.

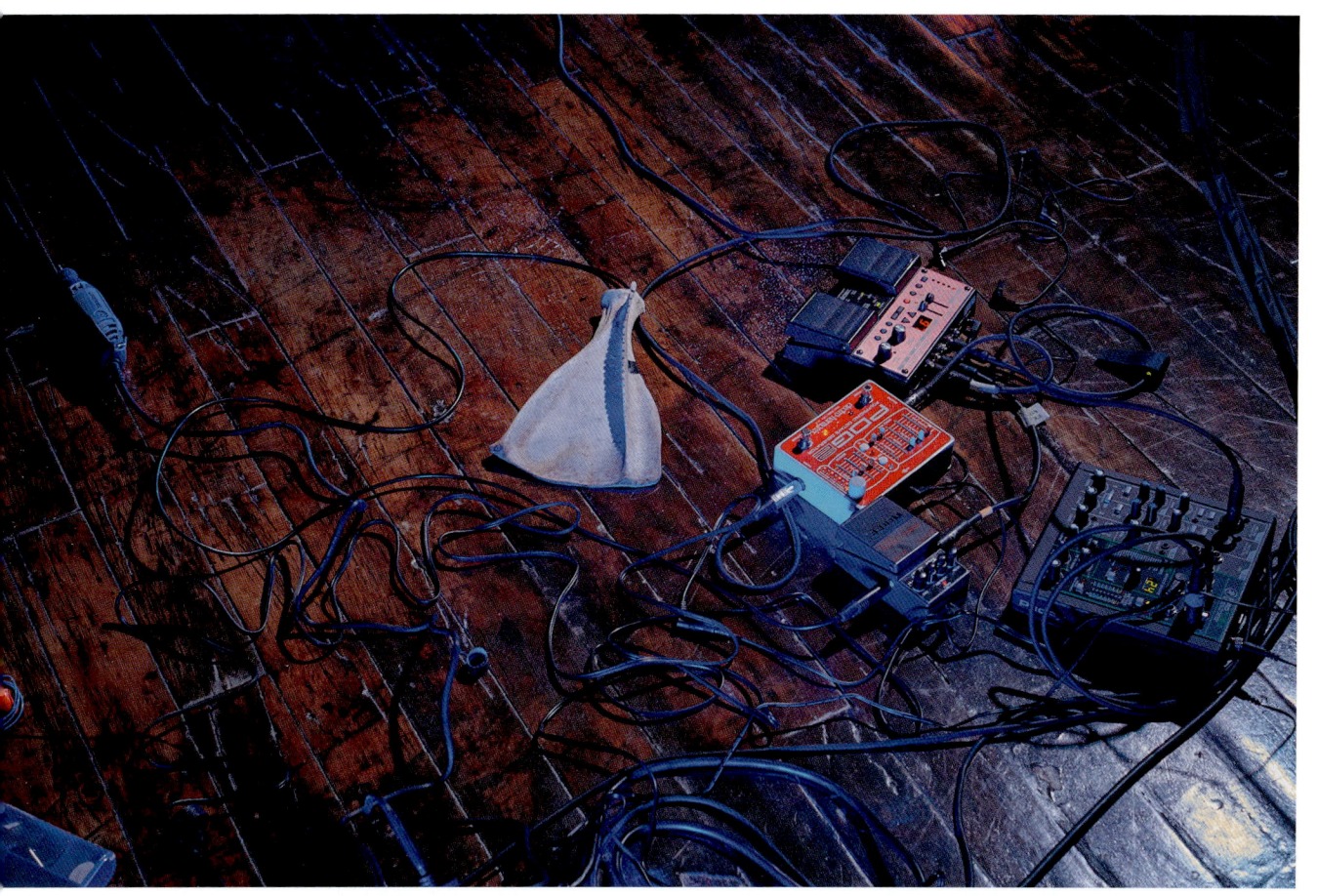

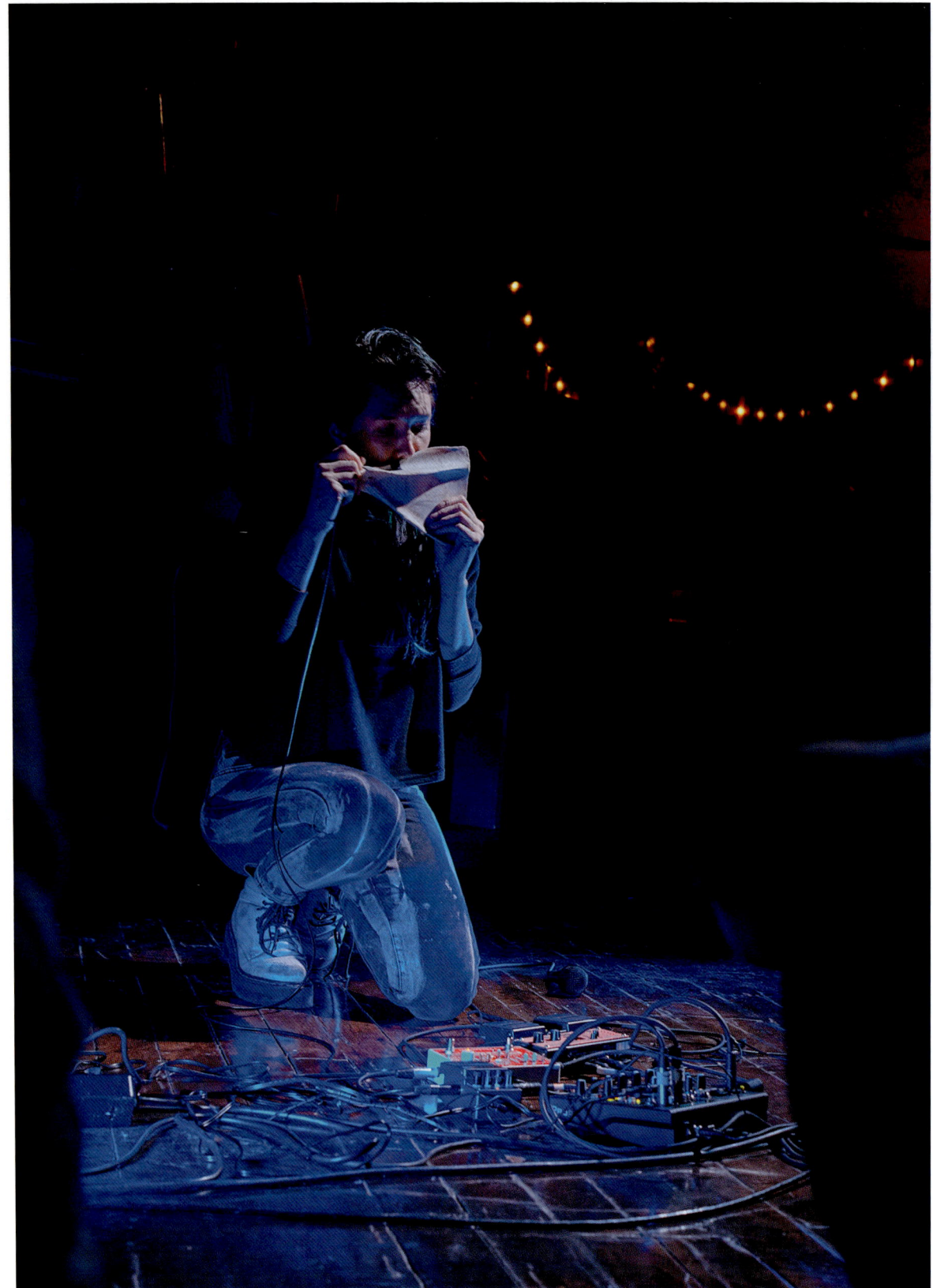

Jeneen Frei Njootli, *Bone meets blade, sonified calling out. who what will utter back*, 2023. Performance at Production Club, Los Angeles, 2023. Photographs by Gabriel Sweet.

DAVID HAINES

JOYCE HINTERDING

David Haines and Joyce Hinterding are individually and together renowned for innovative explorations at the intersection of art, science, and technology. Hailing from the Blue Mountains of Australia, on Darug and Gundengurra country, the duo has carved a distinctive niche in the art world by challenging conventional notions of perception and reality. Hinterding's work often explores the natural world's unseen forces; tapping into electromagnetic phenomena and frequencies, she engages with the invisible energies surrounding us to create sensory experiences. Haines engages photography, sculpture, video, and the olfactory as means to explore how hallucinatory states intersect with the physical environment.

Haines and Hinterding's collaborative works often invite viewers to contemplate how the material and immaterial realms coincide. Their *Telepathy* (2008–ongoing) is an anechoic chamber offering solitude and contemplation. The windowless, wedge-shaped structure blocks out external sound and light, and the hundreds of acoustic foam pyramids that line its interior absorb the energy and convert internal vibrations into heat, enabling participants to focus solely on their own bodily experience. Inside *Telepathy*, the body simultaneously functions as a primary oscillator and sensor. <

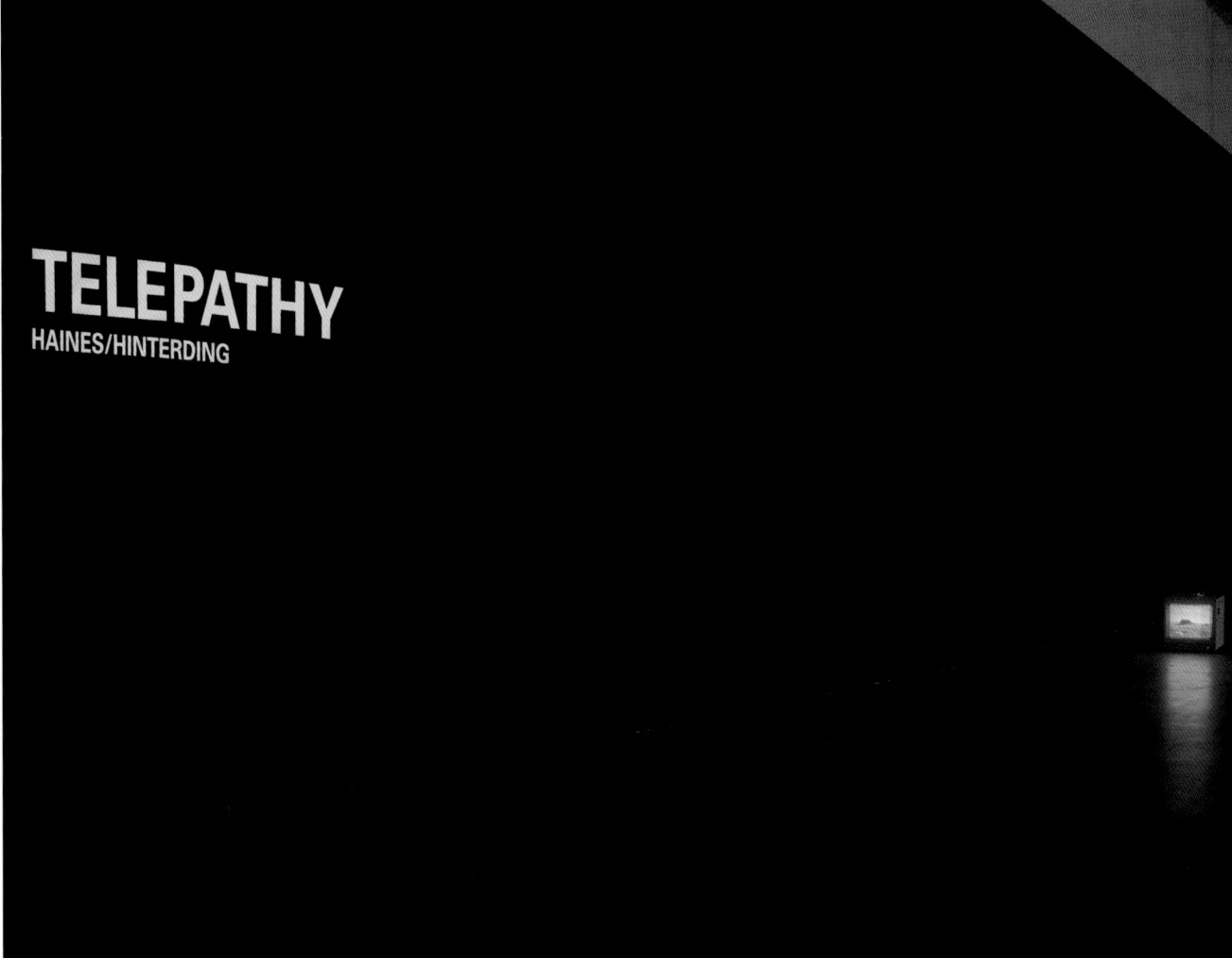

TELEPATHY
HAINES/HINTERDING

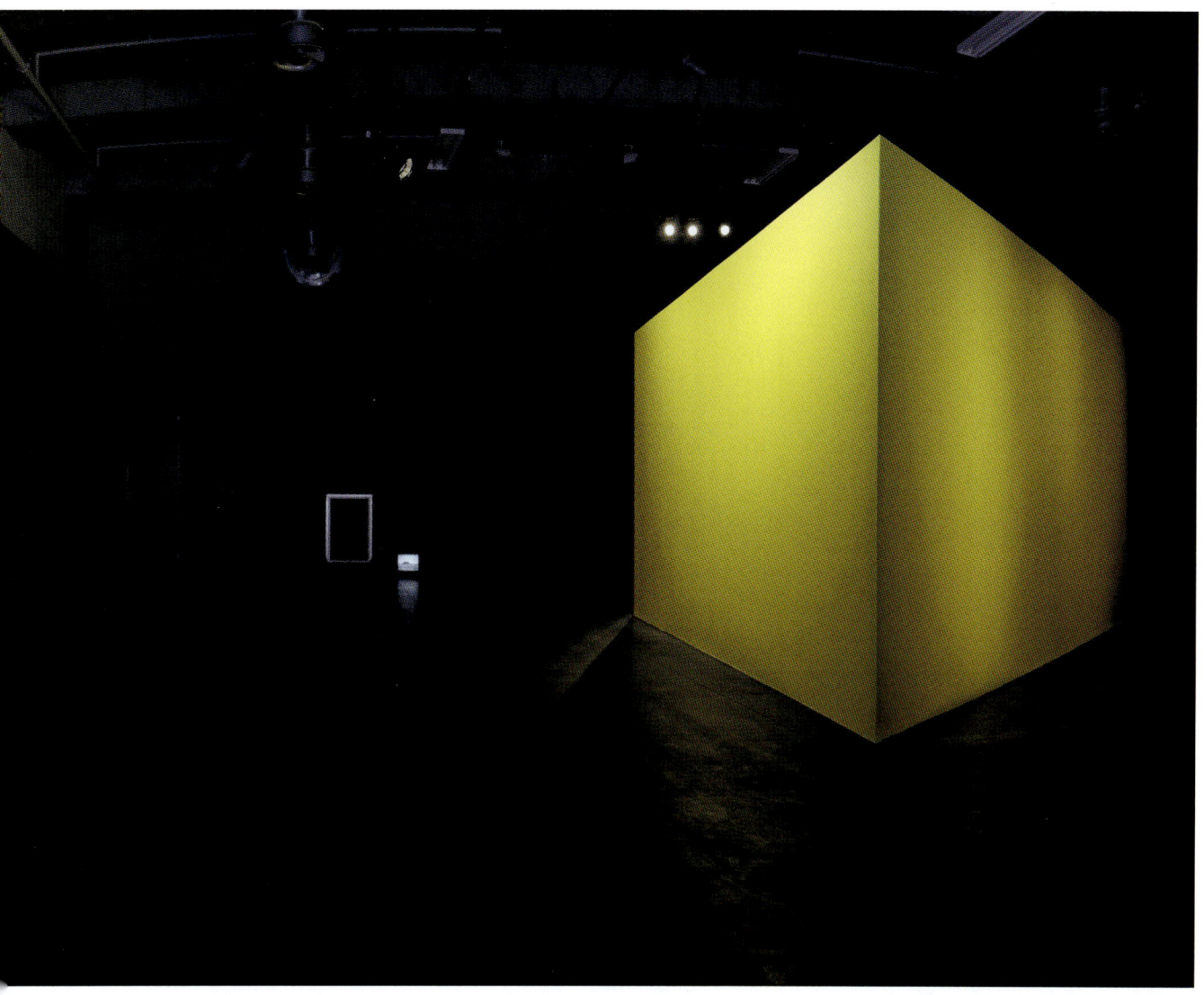

David Haines and Joyce Hinterding,
Telepathy, 2008 – ongoing. Photograph by
Michael Myers, courtesy the artists.

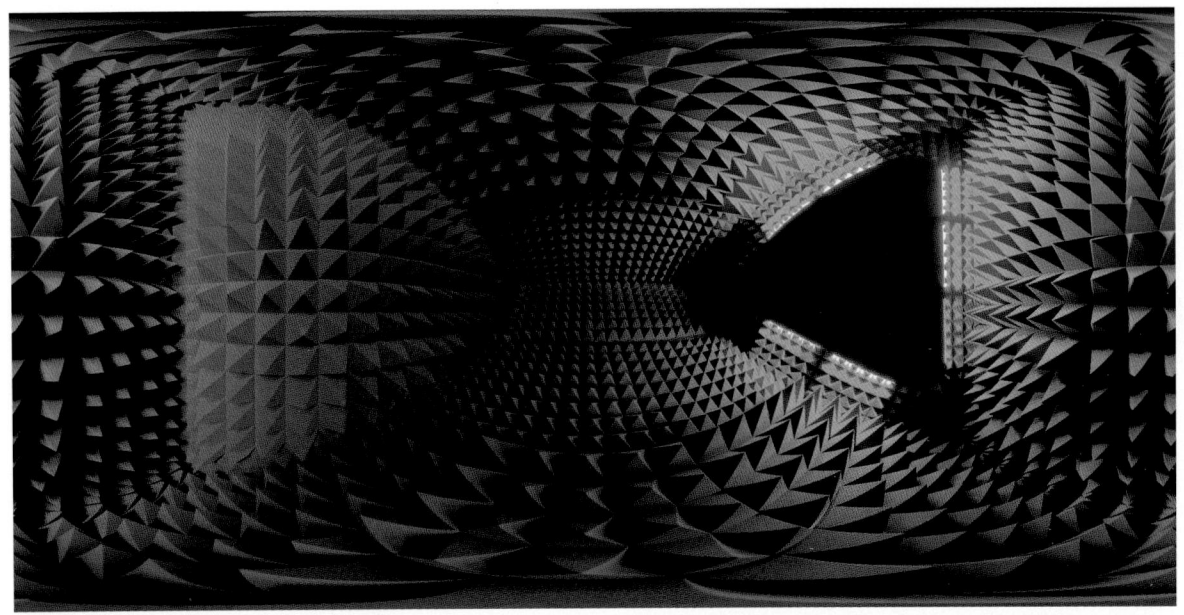

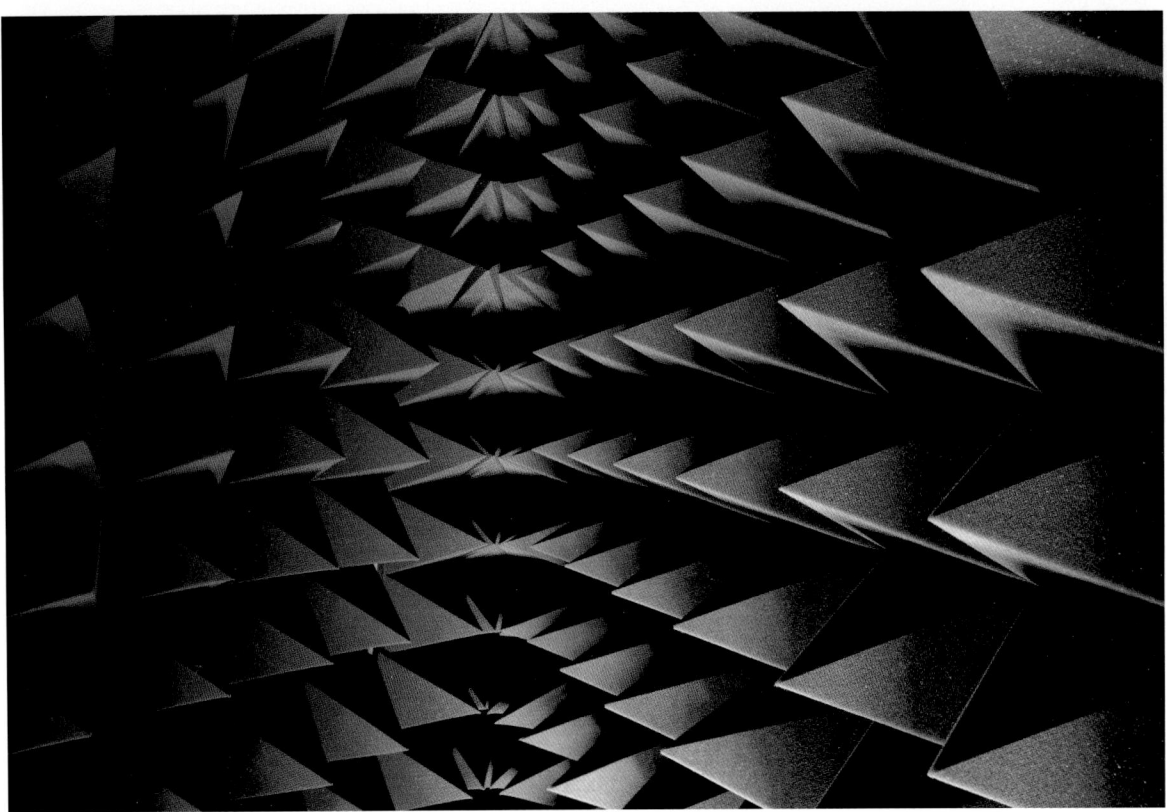

David Haines and Joyce Hinterding,
Telepathy, 2008 – ongoing. Photographs
by Michael Myers, courtesy the artists.

VIRGINIA KATZ

In her exploration of the landscape, Virginia Katz focuses on humanity's relationship with the natural environment. Her work reflects a deep concern for the increasingly fragile connection we share with our planet and emphasizes themes of decay and rejuvenation. Katz's art begins as a metaphor for coexistence as humanity and the earth enter a critical transitional phase. Since 2010, she has created relief paintings and mixed-media monoprints that incorporate found elements such as leaves, soil, vines, tree bark, and Styrofoam. For her 2017 Intervention series, she further questioned and dissolved boundaries between the natural and human made by integrating facsimiles of these elements she made from acrylic paints into outdoor spaces.

Katz created *On-Shore Flow* and *Off-Shore Flow* (both 2008), included in *Energy Fields: Vibrations of the Pacific*, in collaboration with the Southern California wind. To create the works, Katz attached pens to strings she hung from tree branches. As the force of the wind caused the branches to move, the wind's energy was registered via pen marks on large-format paper placed on the ground below. These works represent an energetic exchange of the vibrational environment with human creativity, offering a tangible record of the invisible forces that shape our world. <

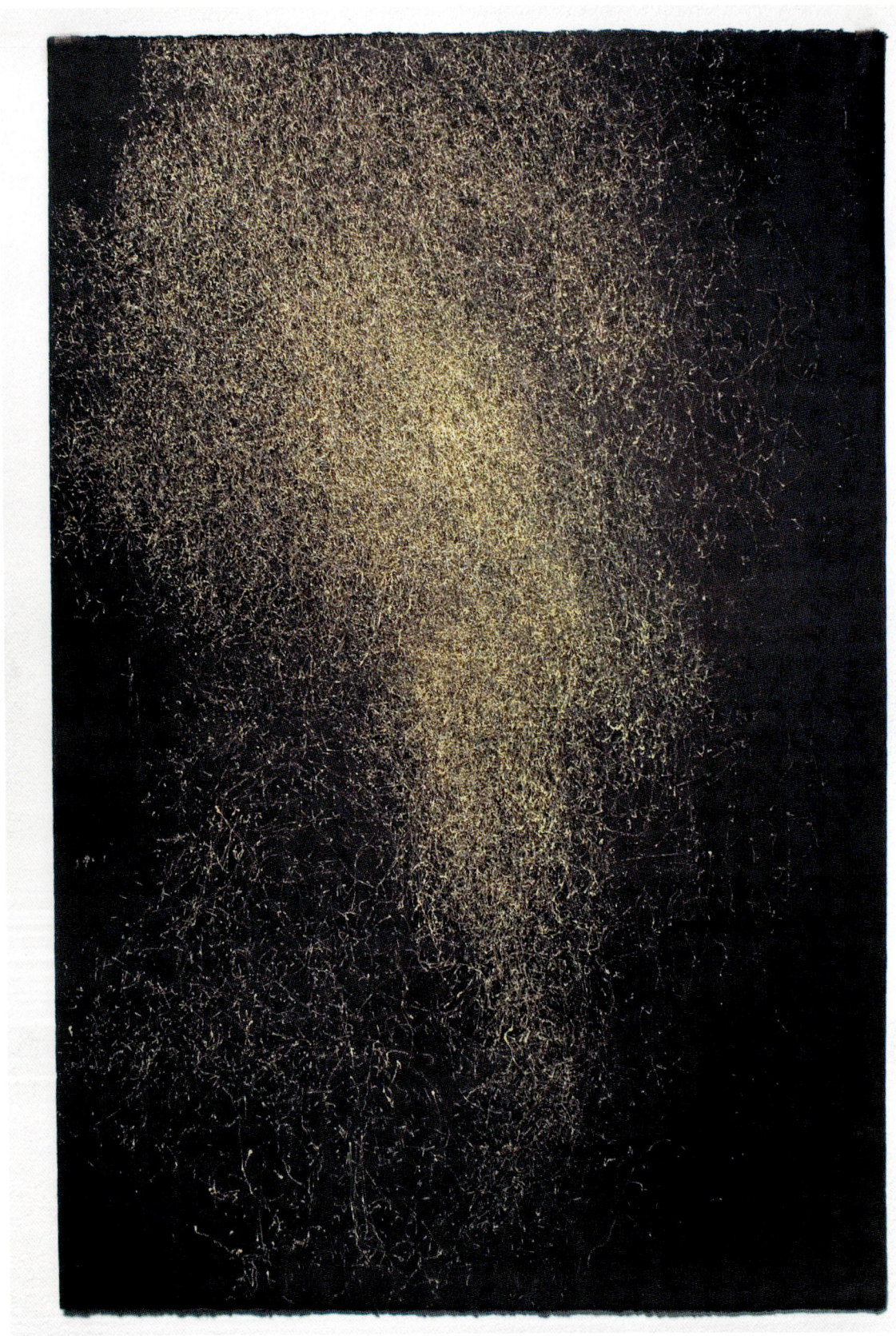

Virginia Katz, *WIND, Off-Shore Flow, 10 Hours of Observation,*
Gold and Copper, 10/07/08, 2008.

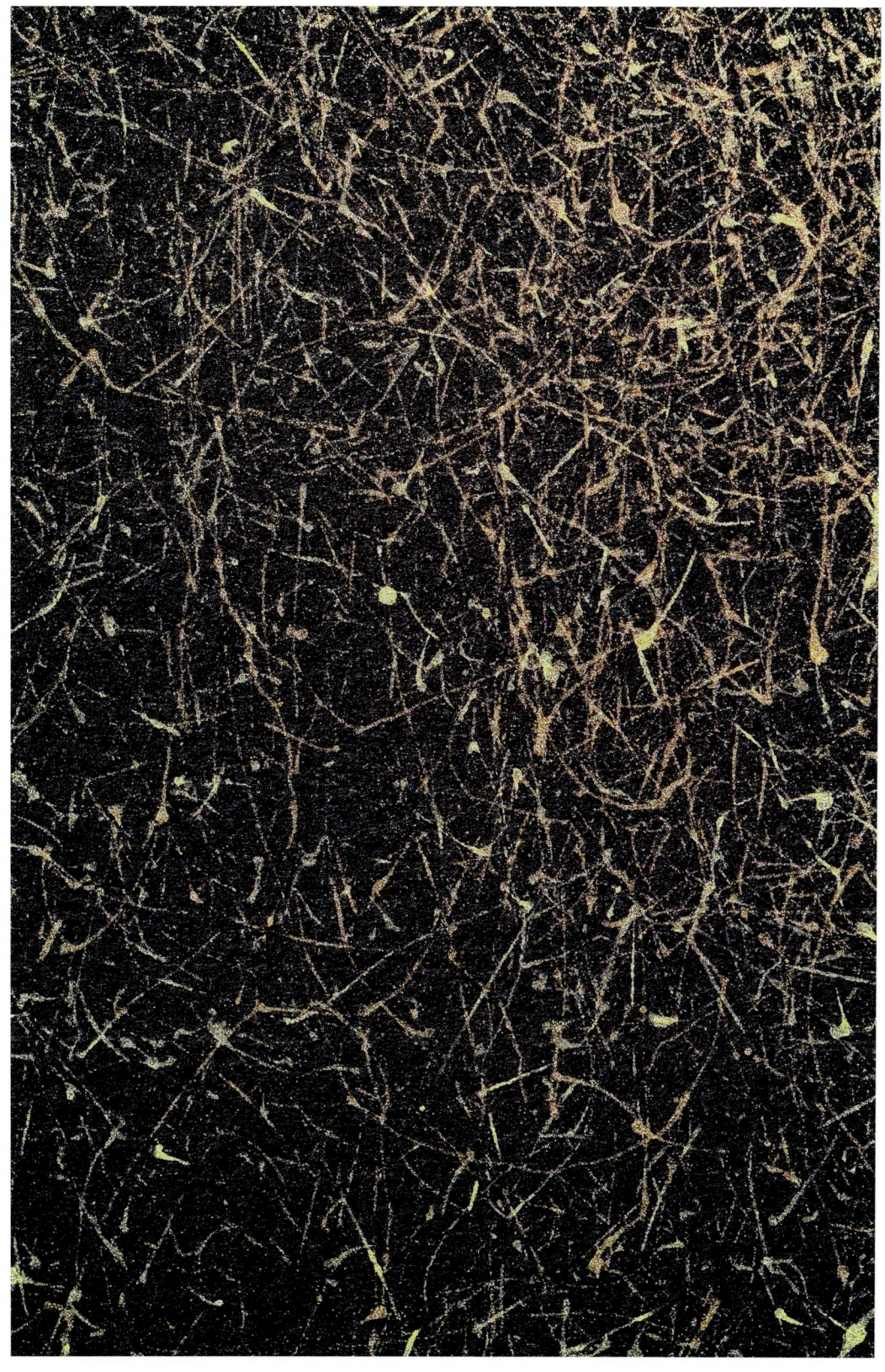

Virginia Katz, detail of *WIND, Off-Shore Flow, 10 Hours of Observation, Gold and Copper, 10/07/08*, 2008.

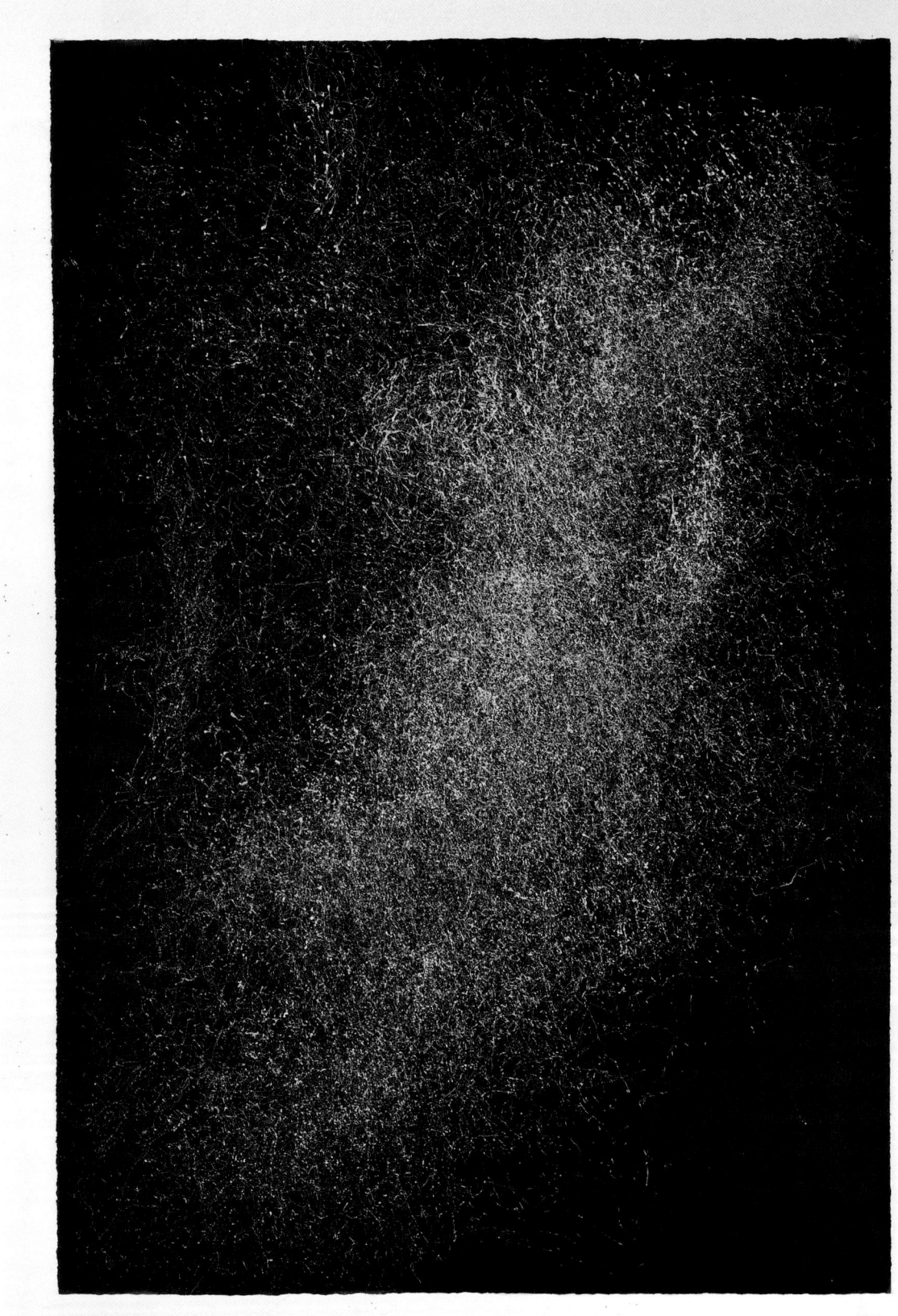

Virginia Katz, *WIND, On-Shore Flow, 7 Hours of Observation,*
Green and Blue, 3/28/08, 2008.

MALENA
SZLAM

The vibrations of the earth and their residual materiality has been a topic of research and deep fascination for Chilean-born, Canada-based artist and filmmaker Malena Szlam. Her work explores the echoing aftermath of tectonic shifts by focusing on the variegated forms, textures, and materials of the landscape. She brings an intensity to her observations, often layering perspective and proportions within a single frame, revealing what she describes as "lyrical approximations of the natural world." As such, her artistic process is an evolving meditation on and through the spaces and environments she encounters.

Szlam's film *Altiplano* (2018), included in *Energy Fields: Vibrations of the Pacific*, was shot in the Andean Mountains on the traditional lands of the Atacameño, Aymara, and Calchaquí-Diaguita in Northern Chile and Northwest Argentina. The work imagines how geographic dynamism, the results of volcanoes, tectonic plate movements, and other formative ruptures, has shaped the world that surrounds us. Traversing diverse landscapes—from vast lakes to salt flats to volcanic deserts—etched and molded by energies emanating from the Earth's crust, *Altiplano* speaks to vibration as a temporal force and a timeless element. <

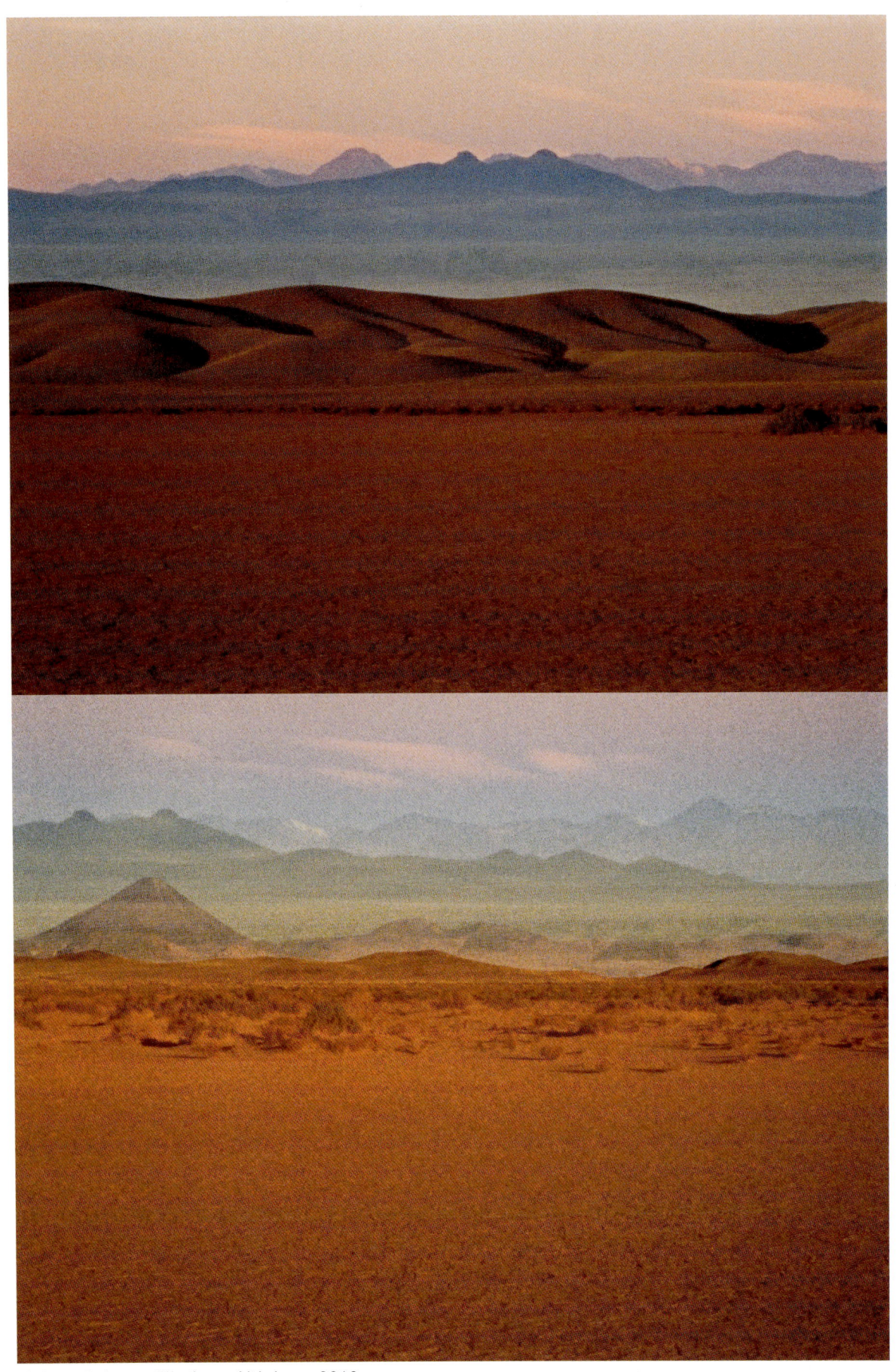

Malena Szlam, stills from *Altiplano*, 2018.
Photographs courtesy the artist.

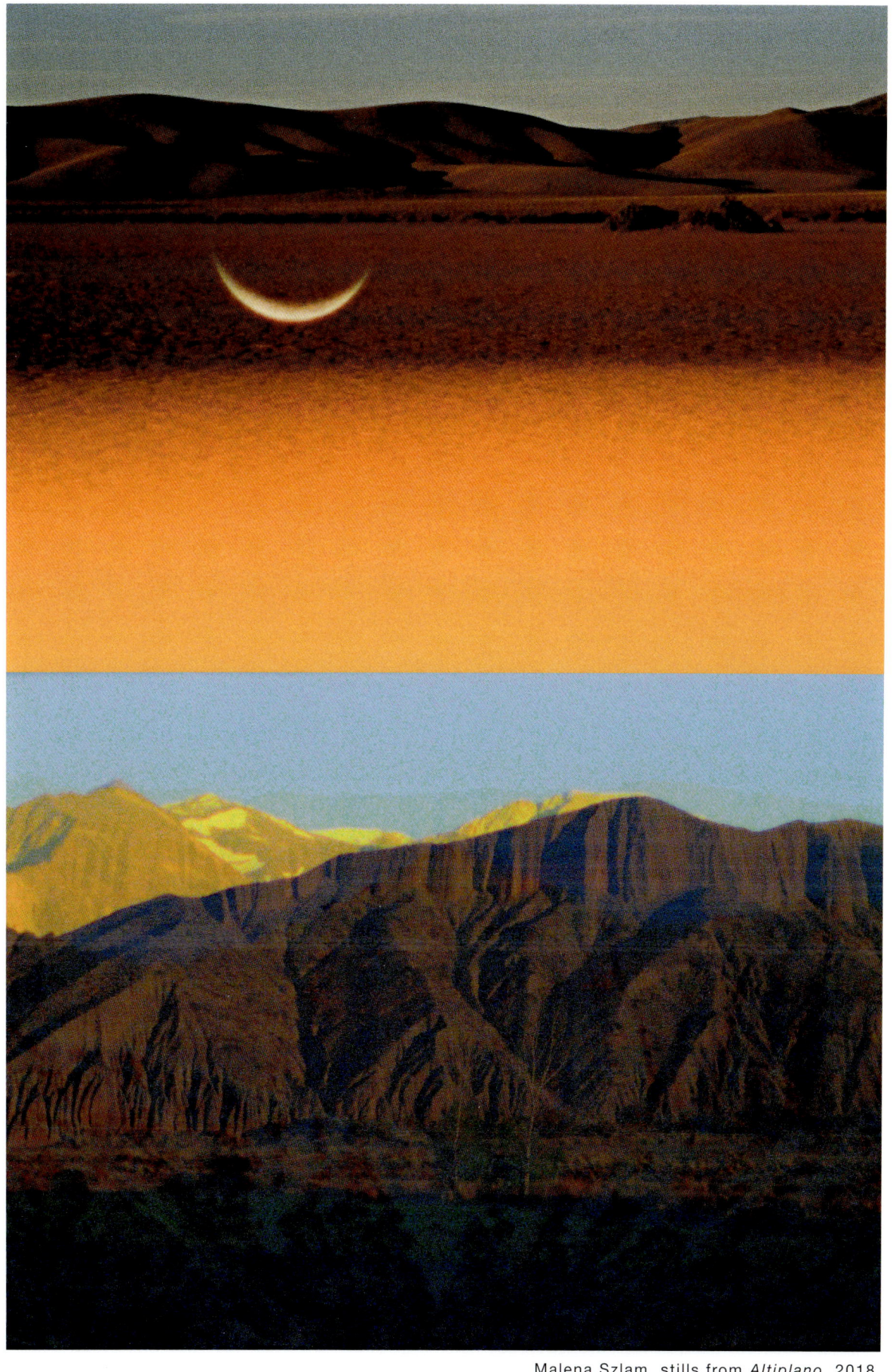

Malena Szlam, stills from *Altiplano*, 2018.
Photographs courtesy the artist.

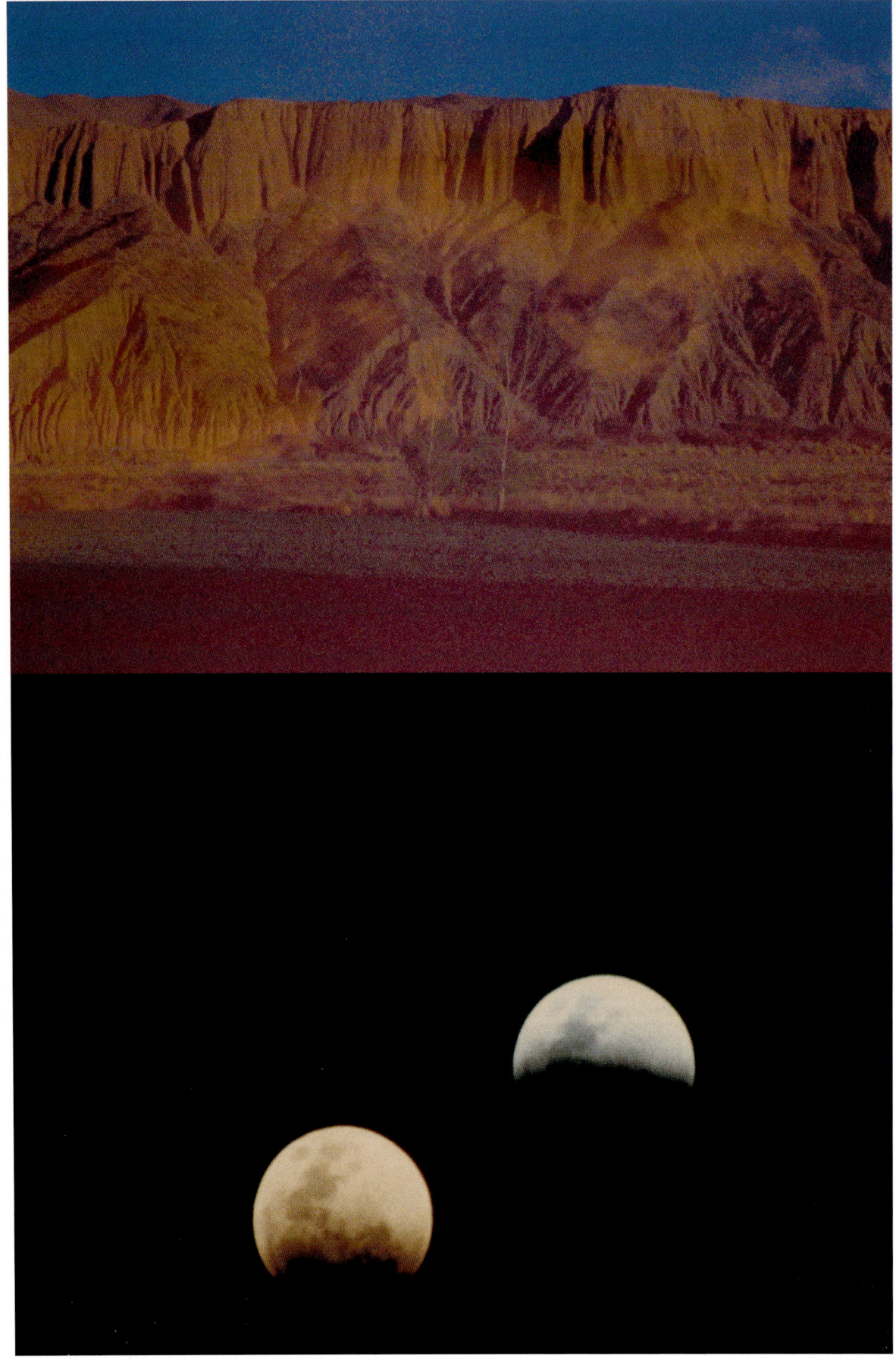

Malena Szlam, stills from *Altiplano*, 2018.
Photographs courtesy the artist.

ELLEN FULLMAN

For *Energy Fields, Vibrations of the Pacific*, Northern California–based sound artists Ellen Fullman and Theresa Wong present *Soundless* (2023), a four-channel installation exploring the qualities of string resonance through Wong's cello and electric guitar and Fullman's unique musical apparatus, the Long String Instrument. Fullman began developing her Long String Instrument in 1980, after graduating with a BFA in sculpture from Kansas City Art Institute. The instrument comprises two sets of twenty stainless steel and bronze strings suspended horizontally across forty-five feet, attached on one side to a wooden resonator. Enough space is left between the two sets of strings so that Fullman can move up and down the length of the instrument to play it, moderating the volume and pitch of the strings' vibrations with her hands. Their collaborative performance is presented in partnership with the Museum of Contemporary Art, Los Angeles, as part of Wonmi's WAREHOUSE Programs. <

THERESA WONG

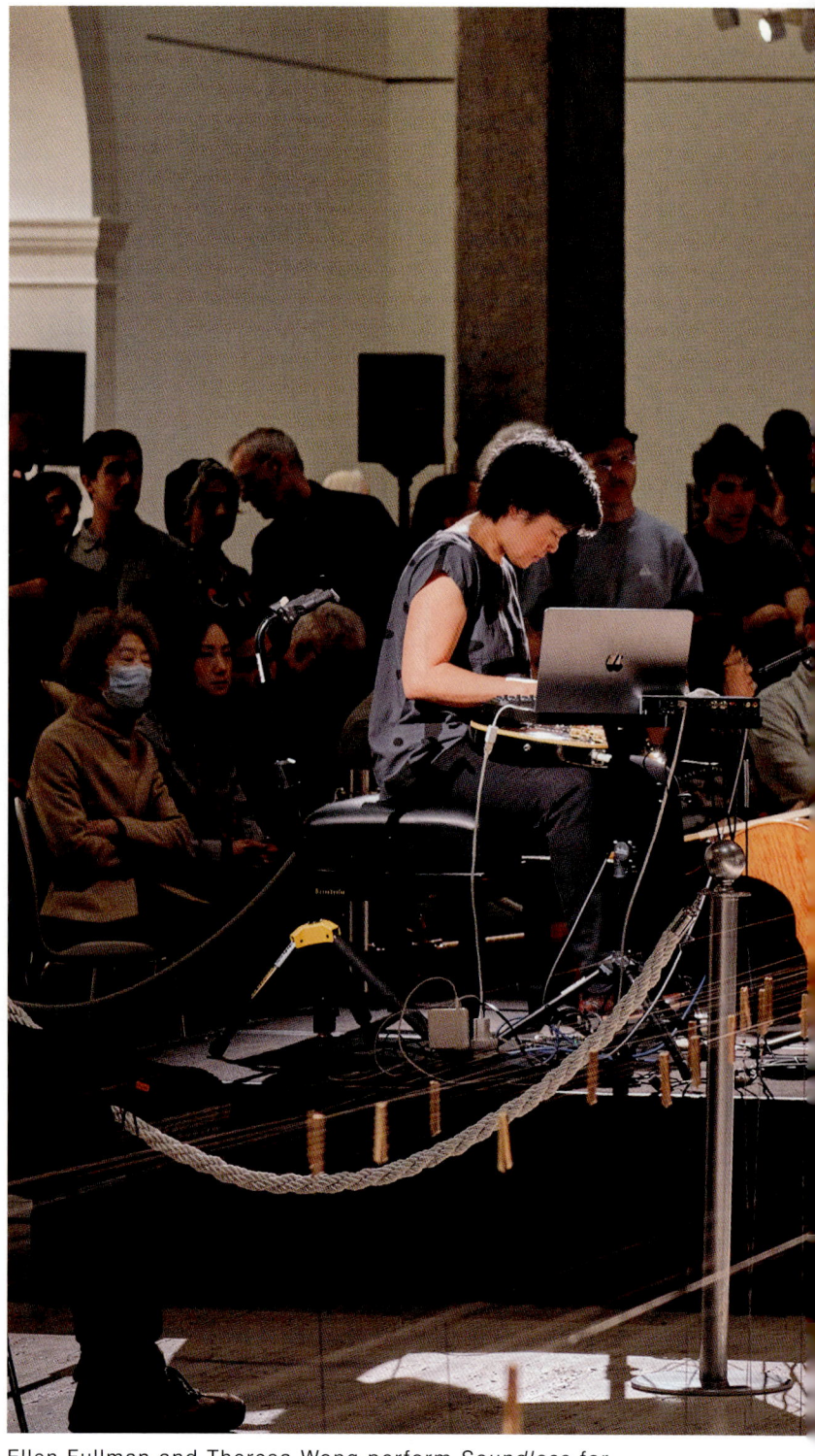

Ellen Fullman and Theresa Wong perform *Soundless* for the Volume festival at the Art Gallery of New South Wales, Sydney, 2023.

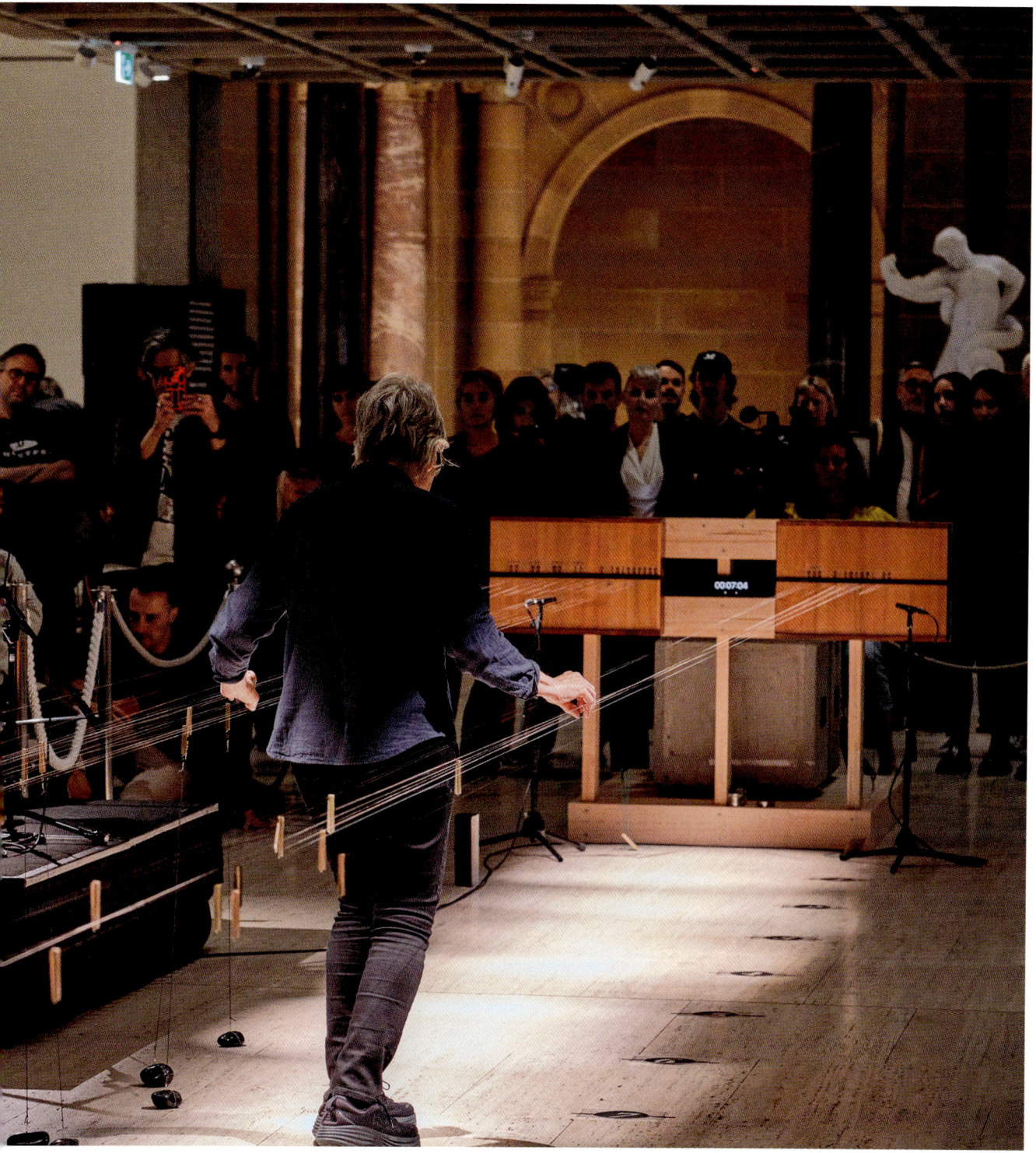

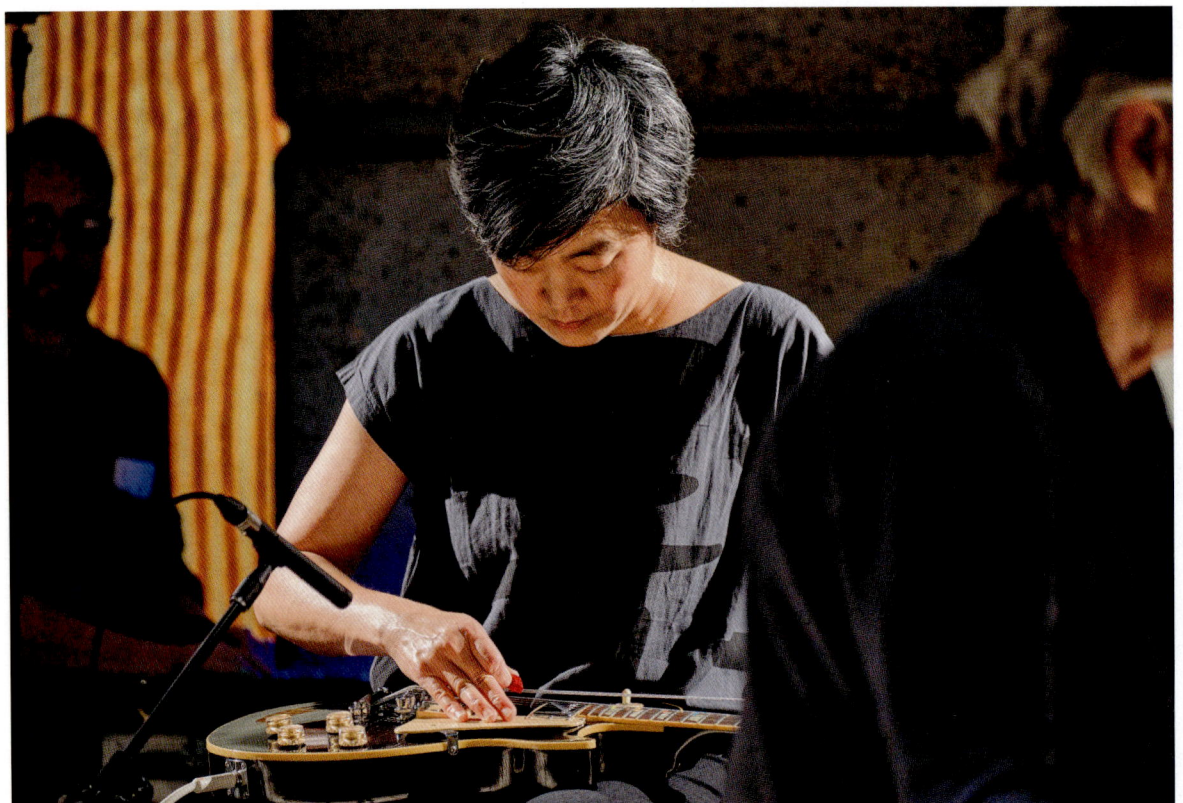

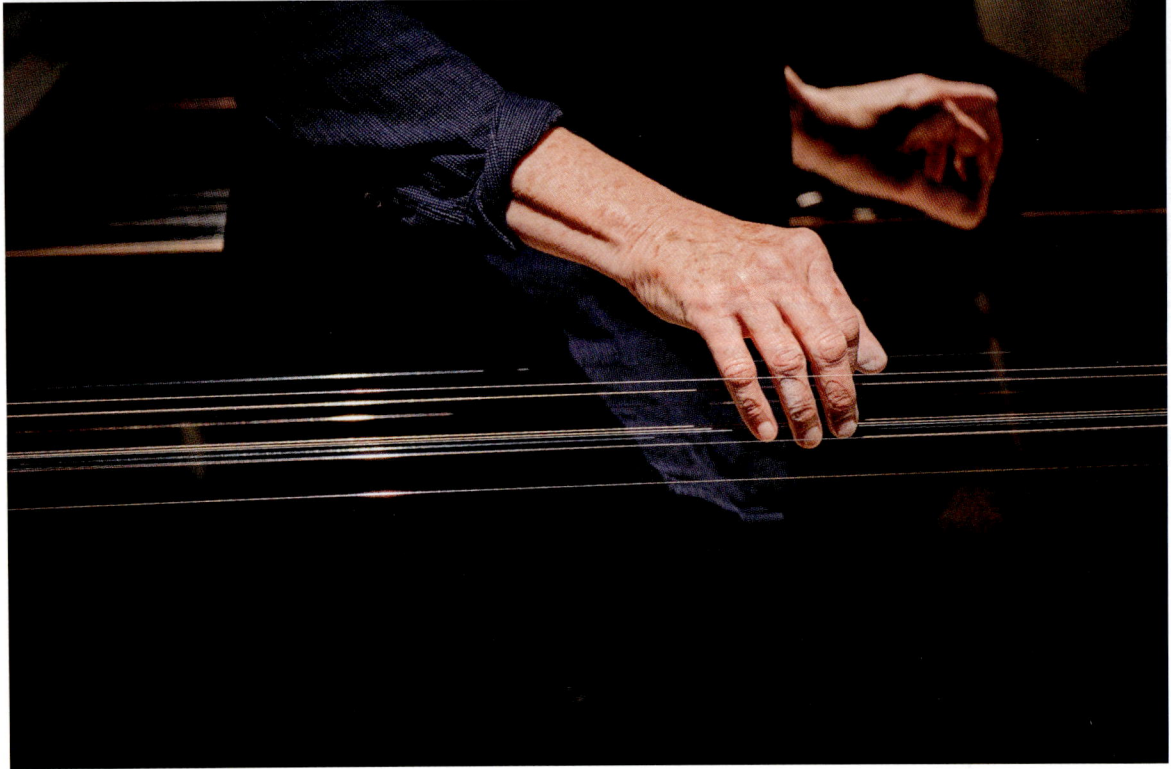

Ellen Fullman and Theresa Wong perform *Soundless*
for the Volume festival at the Art Gallery of New South
Wales, Sydney, 2023. Photographs © Art Gallery of
New South Wales.

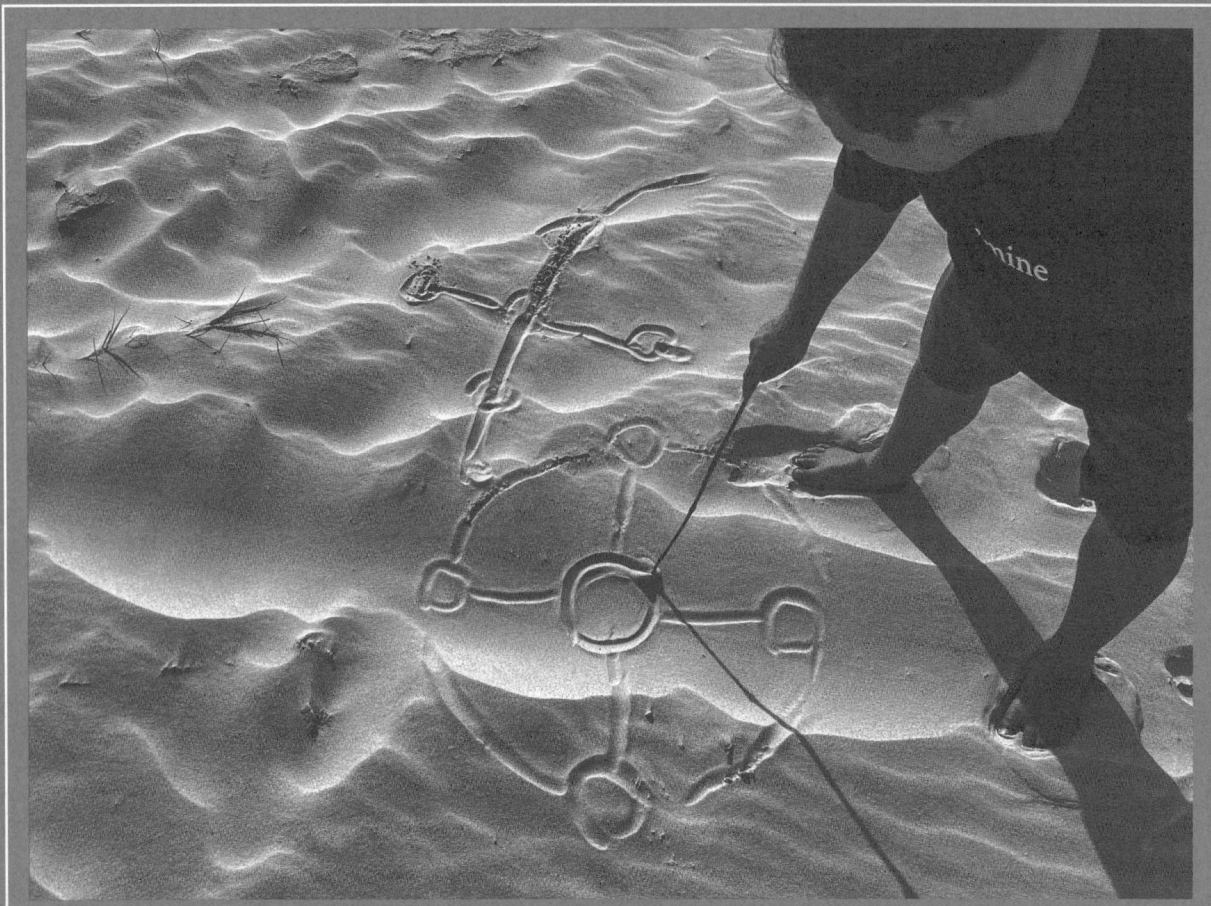

Kyle Slabb, *A starting place of sense, binungal*, 2024. The artist showing the interconnections of land and sky stories at Booningbah (Fingal Head Beach, Australia). Photograph by Lawrence English.

LAWRENCE ENGLISH

in conversation with

August 5, 2023

KYLE SLABB

LAWRENCE ENGLISH: I want to pick up on some of the ideas we've spoken about over the past few years—specifically, your idea of "binungal." I'm interested in how this way of sensing is used in community, and how it connects you to country more broadly. I think many people imagine listening as a practice they know well, and they know how to do. But binungal, at least from what I understand, refers to something more porous and interconnected. What's the apt translation of it?

KYLE SLABB: In our language, a *binung* is a literal thing—it's your ear. When I talk about binungal, I'm talking about listening, but not listening in the way you might commonly think about. It's recognizing where you are, in the middle, but also that you're just one of many middles that stretch outward and interconnect. We use circles as a way of describing many things, and binungal can be characterized this way, as a series of circles that represent the many realms that we might be listening and sensing into.

The first realm is simply the physical. When you're at the center of those rings, you have different spheres of sound or spheres of vibration that come to you. They pass through you, and how you interpret them is dependent on how things move through you, how you are placed, and what you are seeking to experience in that moment. That's the basis of the whole system; it's about all these conditions being experienced together.

In the physical, we can hear what's going between me and you as the center. And we can hear other things around us—some of the birds and some of the things in the house next to where we're sitting now. Further out, we can hear the plane flying past and hear the highway up there, and the ocean out there. In a very simple way, it's about understanding these spheres and what's moving through them as sound and vibration. It's relating how things—physical elements and materials—fit into the relationship we have to a place. Our people think of it as having three layers or three voices that are always changing, and we have to be aware of that as it as it is happening around us.

In your mind, you have a second realm of spheres. There are things that come from inside of you into the realm of your mind. You have things you generate there—we could almost call that a type of self-talk—and things that come in from the outside, but also from within.

You still listen, but the way that you understand sensing might be different. For example, you build an understanding of something from within while taking in a sense of the outside. In this realm, there's a whole other process of sensing that happens. It's outside of time, but also in time. It's about being deeper in listening or just in being more broadly.

LE: So, each of these circles exists around or inside the body, while recognizing the body is situated in a place?

KS: You could say that.

LE: They need to be meshed, though, like the circles you described; everything is overlapping and connecting, and you can't really have one without the other.

KS: Yes, everything is connected for us, and you can't just pull out one way of listening or sensing without the others being part of that. It's deep in that way.

LE: How does the third realm overlay, according to the concept of binungal?

KS: The third realm is what you might understand as an inner spiritual realm. It's not truly inner, though, because it's all around us, physical and nonphysical at the same moment. Aboriginal people sometimes talk about the gut or the heart: it's at the very core of our being. The voice that comes from out there, from way beyond where we can see or sense with our bodies, or from outside of this place, or even from somewhere within: that's all part of this realm. There are another three voices on that plane. It's tapping into something that is not exactly physically present—it's beyond that first realm of binungal. It's how we listen into country and to ancestors, and totems, too. When we feel something from country, we learn to sense that. We come to know what those vibrations, feelings, mean for us.

LE: I think about this as an affective relation—that's the closest reference I can think of for what you're describing: something that's sensed in a very embodied and embedded way. For many of us, though, the understanding of that sensing is something that is not a given but asks us to apply ourselves in time to understanding it, or at least try to. It's about making ourselves available, I suppose.

KS: It is that, but it's also something more than that for us because of our connections here. The three realms that exist for binungal are always existing simultaneously, and within each realm there are three voices, so there's nine different ways that sound or vibration or frequency in whatever realm can come to you.

LE: I imagine that's not something you just arrive at, in terms of being able to appreciate all that information and knowledge.

KS: Those different realms of listening within your mind—I think people don't really understand that. If you're constantly solving between those three voices in your mind, and you don't understand that you even have three voices in your mind—it's like turning the radio dial nonstop. You're just hearing bits of everything, which doesn't make sense a lot. I think even today, approaches to mental health haven't considered some of the cultural understandings of the mind—whose voice, what voice, how we interpret that, and how that connects us to the physical realm. That's the ability that our people have that was always a part of our cultural education. We have that understanding, and when there is a blockage or a problem, traditional healers will call on that system to understand something about that blockage. If things aren't right in whatever sphere, it opens ways to resolve that. How are you hearing? How are you interpreting things? It's a comprehensive, integrated system.

LE: Each of the voices in those three realms, they are all deepening one another, then? I wonder how you articulate those voices? Are they all individual or are they acting together?

KS: Both. You can listen to one and isolate one or have them fold over each other. I think a lot of meditation practices help you focus on one at a time. If you're open to the rest, then the connected messages and the connected voice of all of them help you understand things in a more complete way. For us, we're not just seeking out the one meaning, the one layer, the one thing. Instead, we're thinking about those circles and how they interconnect, how they join and link us to country and to each other.

I think a lot of people do understand it, even if they can't completely comprehend it this way; perhaps it's just that they interpret it very differently. People will walk into a space and say, "The atmosphere is really great." We go, "What are you talking about?" They're not talking about the molecules in the sky. They're saying they sense something on a situational level. Is it to do with the physical vibration? Is it to do with their mind? Is it to do with their spirit? They'll talk about feeling. What is feeling, and how do they interpret feeling?

For us, some of these ideas are held within binungal. When these different voices are connected, it gives you that complete picture of how you're sensing the things around you. An atmosphere can be experienced within just one of the realms, for example, and in some ways, you can create an atmosphere within one of these spheres. You need to be able to understand how you are creating it, though, and understand that process through sound and through other sensory stimulation. Binungal is our way to navigate those ideas. It's our way to take in the entirety of what is around us and accept there are things that are outside of the ways people might usually think about something like listening. It must go deeper, then it's down to how people interpret it and understand that process.

LE: That idea of sensing is interesting, as I feel there's a gap, or at least some distance, between listening and hearing. Perhaps there's a separation of intent that exists between emotions and affect. Here I am thinking of affect as that sub-emotional unknown, or even unknowable perception or feeling of the things around us.

KS: The sense.

LE: Yeah, exactly. Is being available to that affect or that thing that's unknowable part of that process?

KS: Yeah. Having a framework that has a space for it is probably one of the most important things. Asking "Where are you in it?" is the next step. Where are you in that space? There are a few things, depending on your interpretation, your inherent perception of the world, that'll effect the way you see yourself in that space and how you interpret it. I think a conscious availability to it will make a big difference.

LE: The thing I remember from very early on when you described it is the idea of listening to country. The idea of being available to that thing that's physically so far away that you can't readily appreciate it because there's listening happening beyond the acoustic. Part of my mind goes to the idea of vibrations that sit outside our ability to hear them, but it's more than that, obviously. There's a connection there, an ability to reach towards it or be available to it. Even though it's potentially unavailable physiologically, there's a spiritual connection or however you might want to describe it. That's the deepest part of that process.

KS: *Gundala* is hearing something, understanding what you're hearing, which happens in the mind, and transferring that understanding to your spirit. That spiritual realm, the butherum in our language, that's the essence of everything. That spiritual realm is really the deepest layer, and it's the structure of everything. It's where all those spheres and circles of understanding are joining and connecting, it's the foundational structure of everything else. Understanding and having a sense, a connection, in that realm is essential to being able to interpret the rest properly. You can have the other senses, but without *butherum*, it's not complete. A whole part of the potential for understanding sound, vibration, everything is lost.

LE: Do the stories become a kind of interpreting device, almost?

KS: Yeah. The stories we tell connect the physical and the spiritual realms. You've got to think of them like complete landscapes. Just as the whole planet is a complete landscape, there's a whole story that completes that realm for us. There's a whole spiritual realm, and we understand that as a complete realm. It's not bits and pieces, it's not just a pop up here or there. Those realms are complete. That means everywhere there's land, there's story. Everywhere there's land and story, there's a spiritual framework. It comes back to the circles I spoke about—they all cross over, there are connections at every level, and for us this is a way of understanding that. It's also a way of recognizing that there's all this stuff that happens which maybe isn't on the surface. It

LE: A foundation for how to sense and a path to understanding, then?

goes deeper, into the earth and into things that aren't just there. But you have to be open to that.

KS: That's right.

LE: With that idea in mind, how do you think about vibration? Obviously, it's a very physical thing in some respects, but it can be more than that.

KS: Vibration is language. Sound is vibration, and how we communicate with each other is vibration. Even when we're looking at the waves out here on the beach, we know the wind transferred energy to the water, and gravity played a role, too. The water is carrying that energy to land. The physical wave doesn't travel 500 miles, it's the energy through the water that travels. Vibration means the earth is talking all the time. Country is speaking all the time, in and through that vibration. For me, it gets back to a connection to that spiritual realm. You could almost say that the vibration is that realm. How you measure it, and how people understand it—that's another story.

LE: Have you found that, across your life, your capacity to be available to that has shifted? Does it become deeper as you begin to understand the possibility of it?

KS: I think anything you give time to will get deeper. I think the more you pay attention and the more you focus on something, you're always going to get more out of it. I don't think it increases, but I think understanding of it does. When you're young, you feel it and you sense it. It comes to you, but you don't understand it. You get as you get older, or you get more of an education, as our people teach it, you understand it more, you do get a deeper and broader perspective on it.

LE: As we sit here, I can't help but reflect on how the sound around us is affecting the sound of your words. The horizon of audition is expanding and contracting in all directions. Not just up and down or forward and backward, but in all directions at once. It feels like this matrix and meshing that you've been speaking to, and we're bodies within those spheres.

KS: For us, things aren't flat, like the perspectives we might think of when we look at the horizon or something like that. I think the Western world has an addiction to linear and two-dimensional perspectives. Our understanding has always been everything, the universe—everything—operates in spheres. Stories are also connecting and interconnecting. In the teaching of our people, there's a big story and you're not the center. That's our first understanding of identity and belonging: there's a big story and you're just a part of it. That gets you outside of yourself. You're not the center of the story, but you've got to find your place in it. <

by **MARCUS HERSE**

"We have detected
gravitational waves!
We did it!"
—David Reitze, Executive Director,
Laser Interferometer Gravitational Wave Observatory (LIGO)[1]

"We've been studying black holes
for so long, that it's easy
to forget that none of us
has actually seen one."
—France Córdova, Director, National Science Foundation[2]

BEYOND PERCEPTION

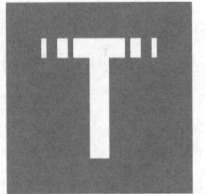he statements above illustrate ways we tend to engage with representations of imperceptible phenomena. They capture the fervor with which we commemorate the outcomes of scientific exploration, simultaneously highlighting our detachment from the realm of terrestrial and extraterrestrial micro and macro phenomena. Our immediate physical experience is shaped by processes of perception, observation, identification, and representation. Our consciousness, however, extends beyond what our senses directly perceive; it is shaped by information and knowledge we have about forces beyond our immediate grasp. Thus, the richness of our direct sensory experience is deepened by a wealth of vibrations signaling the awe-inspiring abundance of energetic occurrences present in our every day.

Consider, for instance, the vast range of electromagnetic oscillations all around us, most of which elude our direct perception. With our eyes, we can perceive a mere 0.0035 percent of the electromagnetic spectrum as visible light. We sense infrared as heat, but gamma rays, y-rays, x-rays, ultraviolet

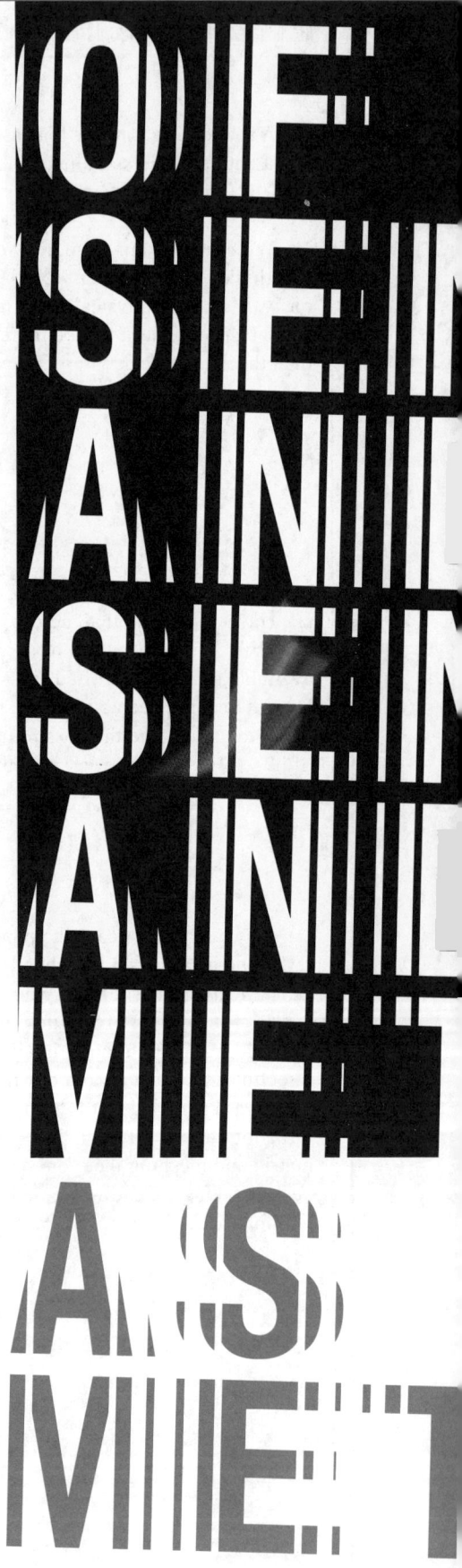

EMBODIED VIBRATIONS:

SES

SORS

APHORS

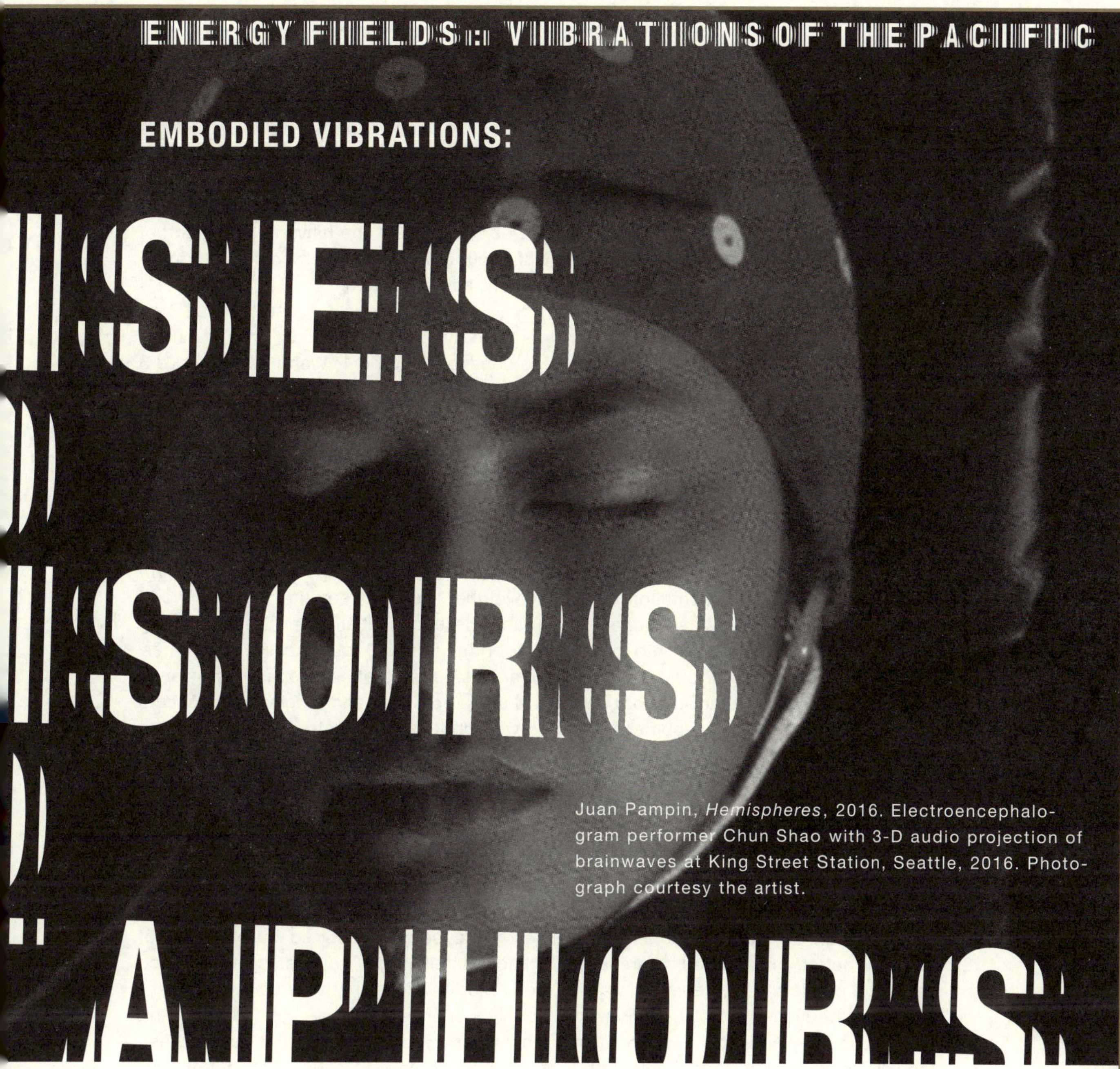

Juan Pampin, *Hemispheres*, 2016. Electroencephalo-gram performer Chun Shao with 3-D audio projection of brainwaves at King Street Station, Seattle, 2016. Photo-graph courtesy the artist.

1. David Reitze, quoted in Tim Radford, "Gravitational Waves: Breakthrough Discovery after a Century of Expectation," *The Guardian*, 11 February 2016, theguardian.com/science/ 2016/feb/11/gravitational-waves-discovery-hailed-as-breakthrough-of-the-century.

2. France Córdova, in "Unveiling First-Ever Image of Black Hole" YouTube video, 1:04:11, 10 April 2019, youtube.com/watch? v=lnJi0Jy692w&t=150s.

A FORCE

waves, microwaves, and radio waves remain imperceptible. Fortunately, we have scientific instruments to serve as indispensable mediators and translators, enabling access to phenomena that necessitate transposition, mapping, reduction, or magnification to render them intelligible within the "limited" frame of our human cognition. These unsensual data, when passed through instruments of mediation, emerge as materials sensorially available to us.

We humans are specialists in making physically absent things present, and we have a great capacity to render the abstract tangible through language and the nonmaterial culture we have developed. Employing metaphors, symbols, and narratives, we traverse the chasm between the concealed and the cognizable, engendering a profound nexus with the enigmatic dimensions of our existence. In the sciences, representation functions as instrumental conduit, permitting the extension of our cognitive faculties into the vibrational frontiers of the world. By conceiving and formulating models that facilitate our perceptions of these concealed phenomena, we strive to grasp them hypothetically and even via prosthetic devices.

Analogous to rendering imperceptible micro and macro worlds tangible in the sciences is the use of art to create awareness. Art manifests the processes, structures, and functions that exist in social and mental realms but remain beyond the scope of externalized, or better, materialized awareness. Art is a kind of public self-awareness in which we direct our attention to initially overlooked aspects of our life-world and encode them in a tangible form. In both the scientific and artistic realms, we employ creative processes to convey abstract systems, relationships, concepts, theories, ideologies, and realities through aesthetic arrangements like diagrams and models or images, objects, installations, and performances. Intriguingly, and typically more ignored in discussions of scientific facts, neither the visualization of (or otherwise sensualized) scientific data nor the artistic representations of facts and circumstances align precisely with their subjects in a straightforward equation. This is so second nature to us that we rarely pay attention to it. A model of

the solar system simplifies vast complexities into something graspable, but it's far from the real cosmic dance. Nobody will claim there is no difference between the two, but the one stands in for the other as a matter of convention. Consider for a moment that almost all you know about the universe and much of what you know about earth comes to you indirectly, as abstraction.

Austrian-British philosopher Karl Popper's theory of mental "worlds" is helpful here. His concept of World 1, World 2, and World 3 categorizes reality into three domains. World 1 represents the physical universe. World 2 encompasses our subjective experiences and consciousness, and World 3 accounts for nonmaterial culture, including language, beliefs, and scientific knowledge. These three worlds interact and shape our understanding of reality. Our representations (World 3) hence form a level that lives between the physical (World 1) and the experiential (World 2), functioning as a bridge in and through reality.[3] This intermediary layer also offers a space for science and art to converge and tangle, which occurs as we ponder the intricate interplay between the seen and the unseen, the tangible and the intangible, the sensed and that which exists beyond our senses.

What if we didn't need World 3 to help us comprehend gravitational waves or electromagnetism and instead could directly sense these forces? Is there a future when we become cyborgs supported by networks of artificial intelligence, advocates of high-tech body modifications that give us a direct sensory experience of these phenomena? And what sensory relations might exist between the body and its prosthetics, between the organic and synthetic materials of the new self? How would it feel, and how would this "feeling like" affect our experience of the world? In "What Is It Like to Be a Bat?," Thomas Nagel argues that there are aspects of conscious experience that are purely subjective: "An organism has conscious mental states

3. Karl Popper and John C. Eccles, *The Self and Its Brain: An Argument for Interaction* (Berlin: Springer-Verlag, 1977).

if and only if there is something that it is like to be that organism—something it is like *for* the organism."[4] We may try to imagine what it is like to be a bat; how it might feel to hang upside down to sleep or flap wings to fly, or what it would be like to use echolocation to navigate the world. But we will only ever be able to approach these experiences from an exteriority, as imagined images and feelings. Similarly, many vibrational energies elude human experience, but we can represent and comprehend them in ways that imagine embodiment and make them relational to us. It is essential to underscore the critical significance of these creative approaches: beyond mere measurement, these are the ways we interrogate, represent, and convey measurements and data sets. It is here that science's link to the arts and to the practices of the artist-researcher become vital. Visual, linguistic, kinetic, and sonic representations stand in lieu of the actual phenomena; objects, models, metaphors, and poetic interpretations map the extrasensory onto our actual senses. Moreover, such artworks not only can explicate the vast expanse of space, mercurial microcosmic worlds, and complex brain activities—beyond these realms they may reveal personal dreams, political utopias, and trans-/post-human musings. Artistic expression, moving in parallel with technological progress, is responsive to paradigmatic shifts and developments. But despite shifting pheno-

types, throughout art history we perceive an enduring undercurrent as a persistent theme: humanity's endless pursuit of the enigmatic unknown and the urge to represent invisible realities.

EMERGING BODIES

It is important to outline, in a radically condensed form, the development of a critical strand of thinking about our access to the world that has influenced scientific discovery over the past few hundred years. During the twentieth century, phenomenology steadily became the focus of modern philosophical inquiry. Edmund Husserl adopted the concept from Georg Hegel's *Phenomenology of Spirit* (1807) and considered it a scientific endeavor guided by the "evidence" derived from the immediate experience of being conscious. According to Husserl, this approach is "intended to provide the fundamental organon for a strictly scientific philosophy and, as a result, to enable a methodical reform of all sciences."[5]

Consciousness, according to Husserl, is always directed towards an object; it is always of or about something. Consciousness is thus an experience of intentionality. It is not itself an object, but rather the qualitative basis through which we experience objects. This perspective seems to cast doubts on objectivism while giving rise to a budding mode of thought that relates to its subject from a first-person standpoint, perceiving it as intricately woven into the fabric of the world and thus forming a foundation for all cognition. In essence, the human perceiver begins to acknowledge themselves as an integral part of the world, shaping the very perception they experience.

Building upon Husserl's work, Martin Heidegger expounded on this idea, recognizing us as intimately intertwined with our surroundings as we contemplate and analyze everyday occurrences. Inextricably linked to a specific social and cultural context and engaging in often-unconscious practical activities like walking, speaking our native language, or using tools, we become an essential component of the environment that in turn shapes our perception of the world.[6] Maurice Merleau-Ponty took this notion further, placing paramount importance on corporeality

---

4. Thomas Nagel, "What Is It Like to Be a Bat?," *The Philosophical Review* 83, no. 4 (October 1974): 436.

5. Edmund Husserl, *Husserliana: Band IX: Phänomenologische Psychologie: Vorlesungen Sommersemester* 1925, available at aerzte-fuer-das-leben.de/pdftexte/overdick-gulden-der-mensch-ist-mehr-banz09.pdf, translation by the author.

6. Martin Heidegger, *Being and Time*, trans. John Macquarrie and Edward Robinson (New York: Harper & Row, 1962), 67–77.

and the profound connection between the body and the world: "The world seen is not 'in' my body, and my body is not 'in' the visible world ultimately: as flesh applied to a flesh, the world neither surrounds it nor is surrounded by it. A participation in and kinship with the visible, the vision neither envelops it nor is enveloped by it definitively."[7] He describes the ambiguity of our existence as an oscillation between consciousness and thingness—each of us is a thing that is aware of itself. Like a hand touching itself, we are the object and the observer of the object, the experiencer and experienced, the perceiver whose perception is also directed towards oneself.

Influenced by treatises on phenomenology by Husserl, Heidegger, and Merleau-Ponty, more recent scholars engaging with cognitive science and the philosophy of consciousness have developed the idea that the entire body and its situatedness in the world form the foundation for consciousness, and that consciousness is shaped by the peculiarities and qualities of the body in and within its environment. These thinkers believe physical engagement with the world directly influences states of consciousness and its formation more broadly. Summarized under the term "embodied cognition," their approach represents a profound shift in cognitive science and encompasses a broad interdisciplinary research field, incorporating philosophy, linguistics, robotics, artificial intelligence, psychology, animal cognition, and neuroscience.

In their 1980 book *Metaphors We Live By*, George Lakoff and Mark Johnson elucidate how language is imbued with metaphors for the body that influence every aspect of our communication, from the most basic to the highly complex.[8] For example, the notion of affection as warmth links physical temperature to emotional closeness, as seen in phrases like "I have warm feelings for you," demonstrating how our bodily experiences shape our understanding of abstract ideas. The uprightness metaphor connects body posture to moral integrity; phrases like "standing up for your beliefs" or "falling from grace" draw on our physical experiences of standing and falling to convey attitudes and judgments. These embodied metaphors, drawn from physical experiences, profoundly influence language and thought, shaping our individual cognition and in turn influencing societal norms, values, and political ideologies.

Extending past linguistics, in *Action's Influence on Thought: The Case of Gesture*, Susan Goldin-Meadow and Sian Beilock argue that "gestures contain detailed perceptual-motor information about the actions they represent, information often not found in the speech that accompanies the gestures... Gesture actively brings action into a speaker's mental representations, and those mental representations then affect behavior—at times more powerfully than the actions on which the gestures are based. Gesture thus has the potential to serve as a unique bridge between action and abstract thought."[9] Gestures are not mere byproducts of thought but are integral to cognitive processes, suggesting that bodily actions can both reflect and shape thinking.

Extrapolating further, Andy Clark and David Chalmers's *The Extended Mind* posits that the mind, or consciousness, is not confined solely to the body but extends into the physical world. In other words, extracorporeal objects, such as our phones, are considered to participate in consciousness: "We advocate for a very different sort of externalism: an active externalism, based on the active role of the environment in driving cognitive processes."[10] They illustrate this point using a case study involving a fictitious character, Otto,

---

7. Maurice Merleau-Ponty, *The Visible and the Invisible* (Evanston, Illinois: Northwestern University Press, 1968), 138.

---

8. George Lakoff and Mark Johnson, *Metaphors We Live By* (Chicago: University of Chicago Press, 1980).

---

9. Susan Goldin-Meadow and Sian Beilock, "Action's Influence on Thought: The Case of Gesture," *Perspectives on Psychological Science* 5, no. 6 (December 2010): 664.

who grapples with mild dementia and relies on a notebook to remember various things, such as directions to the Metropolitan Museum of Art. They contend that for people like Otto, such notebooks serve as information repositories, operating akin to how memory operates for cognitively healthy individuals. Thus, the notebook is an indispensable component of Otto's cognitive process.

As diverse and as specific to finer-grained discourses as these endeavors are, studies of embodied cognition share a common thread: they challenge the dualistic Cartesian model of a rigid division between mind and body, which continues to shape conventional thinking about perceptual interiority and exteriority. Within explorations of embodied cognition, we witness a compelling effort to locate experience as embracing and arising from a web of physical, social, and cultural connections. Recent advances in neuroscience, philosophy, and technology, coupled with a nuanced view of subjectivity, have fueled interest in subjectivist approaches to studying consciousness.[11] These approaches are indicative of an understanding of the world as a collection of processes rather than fixed objects. As physicality becomes central to the way we reframe consciousness, the transient quality of the physical highlights the processual nature of systems interacting in the world. Phenomena of the mind show themselves to be arising from fluid interactions rather than constituting unchanging essences. As a result, consciousness itself is subject to the configurations that give rise to it. The once-clear boundaries between mind and body, which prevailed until recent times philosophically and still extend into our colloquialisms, are now more blurred than ever.

POST-MECHANISTIC PARADIGMS

Scientific method is a mode of investigation that primarily evolved from Western thought, deeply influenced by its Abrahamic origins as well as the rational inquiry promoted during the Enlightenment. It often conceptualizes the universe as a hierarchically-mechanically functioning creation, akin to clockwork, operating according to the rules that govern it. This concept rigidly separates everything within the world, most notably us humans from our environment, proposing binary paradigms and situating understandings in notions of subject-object, God-human, human-nature, and inside-outside, each of which is laden with underlying power dynamics. Of course, this mechanistic worldview presents numerous advantages to unraveling natural phenomena. Its structured approach allows one to comprehend labyrinthine frameworks by dissecting them into simpler constituents, facilitating analysis of their interactions. This methodology also enhances predictability, enabling the development of models and precise calculations grounded in cause-and-effect relationships. Moreover, it drives technological advancements, empowering us to engineer sophisticated machinery and systems. Scientific method has undeniably propelled scientific (and consequently cultural and social) progress, offering an organized way to explore and grasp the universe's functioning intricacies. However, its downside lies in oversimplifying elaborate structures, disregarding their interconnections and emergent properties. A bias toward traditional scientific methods might

10. Andy Clark and David Chalmers, "The Extended Mind," *Analysis* 58, no. 1 (January 1998): 7.

11. Paradoxically, or perhaps quite obviously, the turn to our phenomenal equipment in the Western humanities over the past hundred years is at odds with a strain of scientific study that is increasingly denying direct sensual access: theories precede observations, and observations are corroborated by measurements which are then reified as models. As a result, science has become increasingly abstract throughout the twentieth century. We see here that Husserl's dream of explaining science from the "immediate experience of consciousness," leading to a "methodological reform of all sciences," is as unattainable today as the quantum realm as a vacation spot.

impede a holistic understanding, as it predominantly focuses on individual parts rather than entire systems. Moreover, the approach might limit our perception and interpretation of phenomena operating beyond linear cause-and-effect relationships, effecting our comprehension of intricate biological, ecological, or social networks. On the other hand, a method of inquiry centering the idea of a steadily evolving, decentral and interconnected world and alternative paradigms offers more comprehensive insights into the entanglements of the natural world. In neuroscience and embodied cognition, the brain is understood as part of the dynamic organization of the body that itself is part of and interacting with its environment. This approach emphasizes the influence of a mesh of external systems on consciousness, transcending the limitations of a mechanistic framework and offering a more integrated understanding of consciousness and its relationship with the world. This method of inquiry has also been taken up by artists to pose questions that move beyond the linearity of a mechanistic worldview and suggest an expanded set of relations between self and the world. They make use of an open playing field for differently structured investigations, one that provides creative permission to explore phenomena in a porous and often multidisciplinary way. Our evolving view of the world recognizes the interconnectedness of our society and what we call nature, encouraging comprehensive understanding, ecological ethics, and global responsibility and influencing how we approach science, philosophy, and cooperation across academic fields. In this context, the embodiment of intangible vibrations challenges conventional boundaries of consciousness.

The following sections explore three works by three artists native to the Pacific Rim which contemplate the interiority of perception (the perceiving body) and the models of exteriority that permit us to reach into the enigma of "outsideness." While these works are grounded in technology and scientific methodologies, they wonder: What might be the modes of encounter with vibrational forces? What do different incarnations of these forces lead us to believe about them? How does their embodiment expand and influence

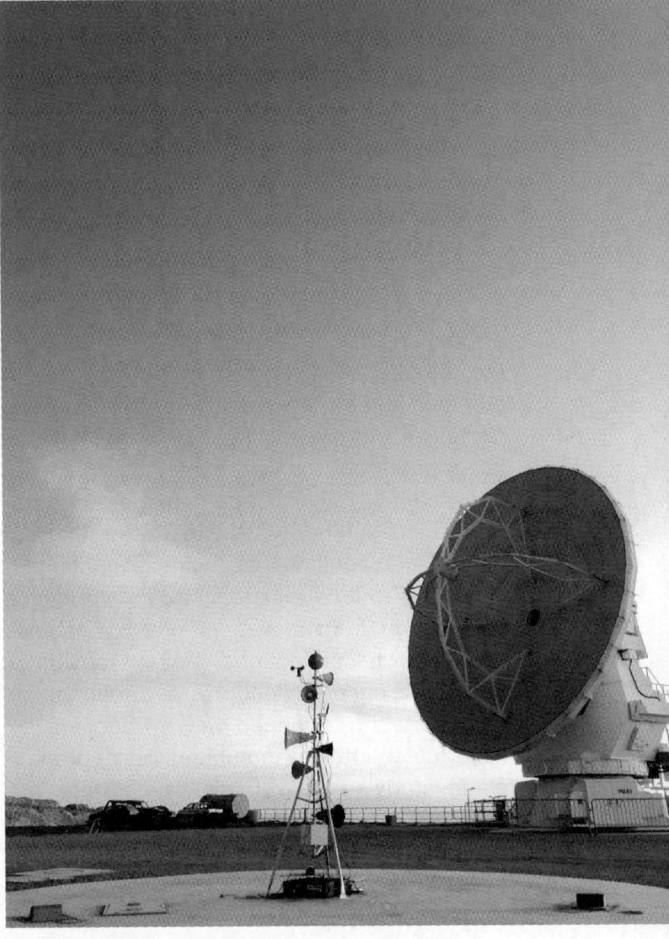

Nicole L'Huillier, *La Paracantora*, 2019. Traveling sonic sculpture installed at the ALMA Observatory in the Atacama Desert, Chile, 2019. Photograph courtesy the artist.

our consciousness? If we take seriously the profound impact our own embodied experience has on what we call our consciousness or mind and the notions of the extended mind theory as laid out by Clark and Chalmers, what does sensualized data mean for our understanding of the world and ourselves in it?

LA PARACANTORA: EMBODIED HYPEROBJECTS

Chilean artist Nicole L'Huillier's sound sculpture *La Paracantora* (2019) delves into the human encounter with invisible natural forces. The work consists of an eight-and-a-half-foot-tall tripod mounted with a multitude of sensors connected via computer and amplifier to six colorful loudspeakers. The sensors measure air pressure, altitude, temperature, acceleration, electromagnetic fields, wind, light, proximity, and vibrations. The speakers, which point in multiple directions, translate the sensor data into sounds. The work has been installed at some of the world's most advanced sites for the exploration of the cosmos, including the European Organization for Nuclear Research in Switzerland and the Paranal Observatory in Chile.

The artwork conveys a message that invites contemplation, directing attention to the immediate surroundings. Implicit within this message is an assertion that the remarkable phenomena represented by *La Paracantora*, albeit intangible, are not distant or theoretical but part of the observer's reality. The work's sensors are set globally, effectively capturing the unique environmental conditions of each location where it is installed. During a 2019 performance at the Large Hadron Collider in Switzerland, *La Paracantora* interpreted the surroundings. Operating at a quarter mile above sea level, the work's altimeter and thermometer generated readings, translating invisible forces into a medley of droning, hissing, and gurgling sounds within the catacomb-like environment. A few months later, at the Atacama Large Millimeter Array Observatory nestled in Chile's Atacama Desert, situated two miles above sea level and exposed to a forty-degree temperature shift between day and night, the work emitted notably different sounds, producing a soft, repetitive melody complemented by gentle undulating tones almost amounting to a musical chord. Subsequently, at the Paranal Observatory, also in the Atacama Desert, *La Paracantora* transitioned between birdlike chirping and serene bubbling reminiscent of a cascading mountain stream.

L'Huillier, who is a musician and experimental composer and holds a PhD in me-

dia arts and sciences from the Massachusetts Institute of Technology, created the audio to represent invisible forces through an intuitive process. The artist has said she was mimicking nature's signals, for instance translating vibrational data into granular percussive sounds or transforming wind patterns into swirling and scraping noises. The work creates different auditory landscapes, each possessing a unique, dynamic character and offering a precise reflection of its genius loci. *La Paracantora* conveys a natural manifestation in an electronic form; it is an instrument played by the forces of nature. As L'Huillier says, "*La Paracantora* is an artifact that helped me to explore CERN from other sensitivities and to engage with the performative conditions of our reality(s) through the sonic dimension. It became a bridge with the unknown, the invisible, and the cosmic mysteries that are being explored there every day. This experiment was for me a way of practicing listening as a means to encounter the multiple realities, temporalities, agencies, and dimensions we inhabit."[12]

La Paracantora is a sonic analog to tomography, a shamanic device through which

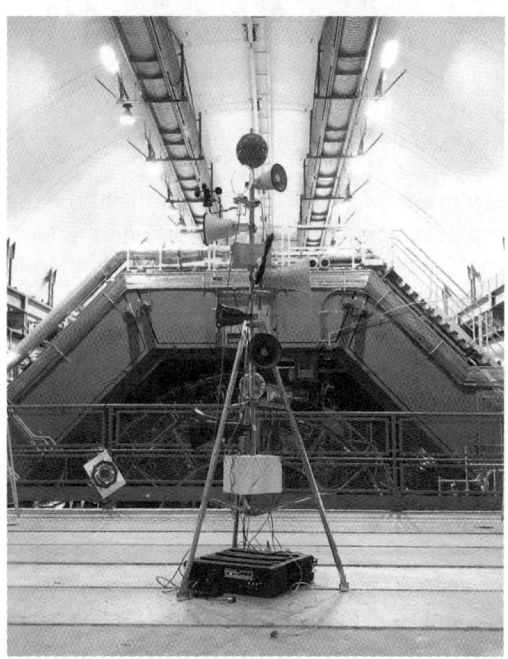

Nicole L'Huillier, *La Paracantora*, 2019. Traveling sonic sculpture installed at the ALICE Detector at the European Organization for Nuclear Research (CERN), Switzerland, 2019. Photograph courtesy the artist.

12. Nicole L'Huillier, quoted in Ana Prendes, "Nicole L'Huillier Pays Homage to Walter Smetak with Her CERN's Sonic Experiments," *Arts at CERN*, arts.cern/article/nicole-lhuillier-pays-homage-walter-smetak-her-cerns-sonic-experiments.

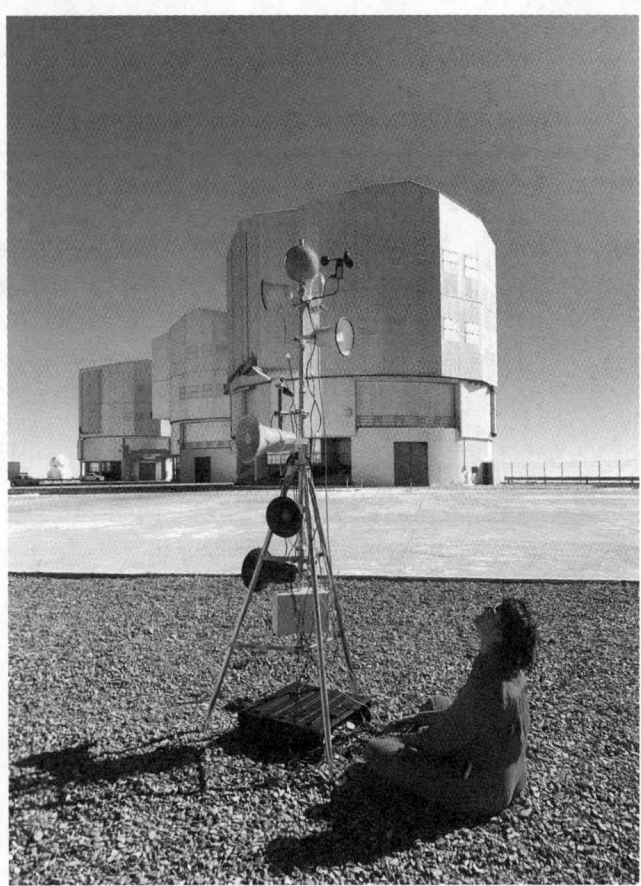

Nicole L'Huillier, *La Paracantora*, 2019. Traveling sonic sculpture installed at Paranal Observatory in the Atacama Desert, Chile, 2019. Photograph by Barbara Nunez.

loudspeakers pointing in different directions signal the work is aimed at a broad public: "The speakers are not earphones. They're not for one, they're for many."

La Paracantora is not an immersive, surround-sound installation emulating the vaporous presence of the detected invisible forces. Instead, it serenades us from a single point. The work thus engages not only in the sonification, but in the personification of nature; it is an eccentric vocalizer, weaving together various forces that become, embodied as *La Paracantora*, music to our ears. The title of the work, "the Parasinger" in English, takes a feminine noun in the Spanish, highlighting the channeling of elemental energies through a female singer as central. Referring to the work as a singer, given the ethereal and shapeless nature of the summoned forces, is a sagacious choice, acknowledging the profound role that singing plays in human experience. According to Danish linguist Otto Jespersen, singing predates our language, and our speech developed from indefinite guttural contact sounds.[15]

Against this backdrop, what is perhaps the most intriguing fact, as beautiful as it is perplexing, is that the scientific staff at CERN embraced L'Huillier's vision and joined in singing with the artwork, humming and chanting along with the work's divestments. Singing, unlike speech, can transcend the need for words. It doesn't rely on metaphors or symbolism to make its presence known; rather, it emerges as a pure vibration that exists in itself, not representing something else.

layers of reality are brought forth. It acts as the proverbial ambassador for Timothy Morton's hyperobjects, structures so extended in time and space (like atmospheric rivers, the ozone layer, or ocean circulation) that we have no sensibilities for them, as we are embedded within these processes.[13]

La Paracantora "could have [...] been a computer and a sound system, that's all you need,"[14] says L'Huiller. This gets to the heart of the problem, namely that invisible or inaudible forces can only play a role if they are heard or seen. The public nature of the sculpture is evident in its appearance: "Bright colors and weird shapes, that's me" says the artist with a smile, making clear in passing that poetry needs no justification. In contrast to purely scientific instruments, *La Paracantora* raises the question of the social function of data, emphasized by the design of the work. The

13. See Timothy Morton, *Hyperobjects: Philosophy and Ecology after the End of the World* (Minneapolis, Minnesota: University of Minnesota Press, 2014).

14. L'Huillier, interview with author, 7 June 2022. Subsequent quotes are from this interview.

15. See Otto Jespersen, *Language: Its Nature, Development and Origin* (London: George Allen and Unwin, 1922).

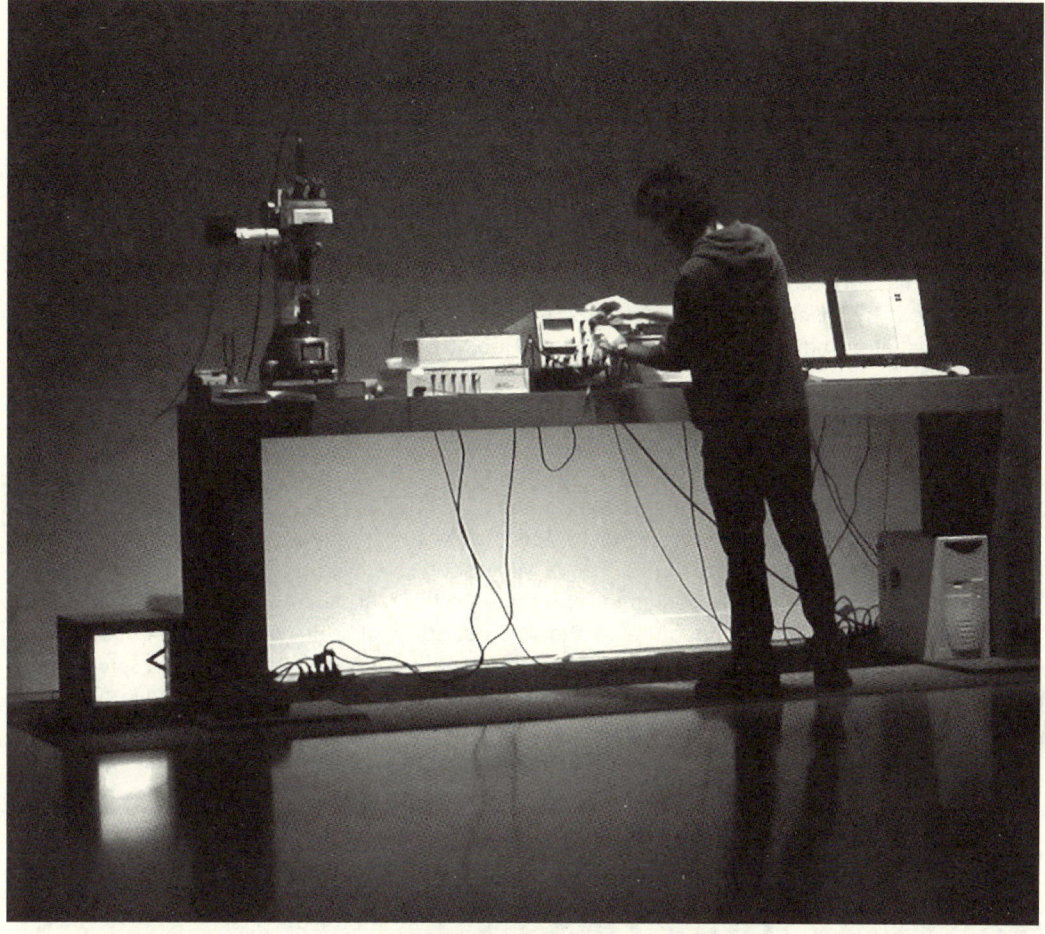

Joel Ong, *Nanovibrancy*, 2011. Site-specific sound installation and performance at John Curtin Gallery, Perth, Australia, 2011. Photograph courtesy the artist.

In this manner, L'Huillier achieves the remarkable feat of bringing dimensions of reality that we can only intellectually experience closer to us, in a human and genuine manner.

La Paracantora is a messenger of the imperceptible and a memorial to transhumanism. One can only speculate about the nature of the communication that occurred at CERN. The episode demonstrates, however, that in order for us to engage with phenomena we must perceive our reflection in them. Human truth and cosmic truth may lie beyond what can be expressed in words—beyond what can be fully grasped intellectually. However, it may not be beyond what can be sung, or to freely paraphrase the lighthearted yet profound Alan Watts, we might be able to express it with that which swings.

NANOVIBRANCY: A MONUMENTAL EARDRUM

Our ears are sensitive instruments that enable us to detect vibrations that move the eardrum by less than one picometer—one thousandth of a nanometer, or about 100 times smaller than a hydrogen atom.[16] Joel Ong's four-hour performance installation *Nanovibrancy* (2011) harnesses the potential of an expanded sensorium and directs it towards exploring its own unique qualities. In this case, the work eavesdrops on the sound of the human eardrum, asking: "What would the eardrum sound like if we were small enough to stand near it?"[17]

16. Wenxuan He, David Kemp, and Tianying Ren, "Timing of the Reticular Lamina and Basilar Membrane Vibration in Living Gerbil Cochleae," *eLife* 7, no. e37625 (2018).

We can hypothesize that if we could shrink to microscopic size, we would be exposed to a bewildering and hazardous microcosmos, the once-familiar world transformed into a colossal and threatening landscape. Air molecules would be densely packed and viscous like honey and oxygen would be scarce, posing severe challenges to breathing. Gravity's grip on us would weaken, making controlled movements difficult. Temperature fluctuations would become extreme, jeopardizing survival. Our mobility would be restricted by seemingly insurmountable obstacles, while our usual communication methods would be virtually useless.[18]

In July 2011, Ong, who is an associate professor for computational arts at York University in Toronto, managed to metaphorically shrink by utilizing a scanning electron microscope and a model tympanic membrane at John Curtin Gallery at Curtin University in Perth, Western Australia. Ong reappropriated the Atomic Force Microscope (AFM), an instrument originally designed for imaging rather than sonification. The architecture of the microscope makes it well suited to capture sounds: Conventional optical microscopes use light to view the sample under investigation, but as the size of the sample decreases, making it bright enough to see becomes increasingly challenging. Moreover, the magnification capability of optical microscopy does not extend beyond two micrometers, which is slightly smaller than the diameter of a string of a spider's

17. Joel Ong, "Nanovibrancy: An Auditory Performance of Nanoscale Resonance," unpublished artist statement available at isea-archives.siggraph.org/wp-content/uploads/2020/06/ISEA2011_333_Joel-Ong.pdf.

18. For detailed speculations on how humans might experience the world at nano scale, see "If You Were Shrunk to Microscopic Size Would You be Able to See Normally? Would You Be Able to See Microscopic Things?," Ask a Mathematician/Ask a Physicist, 2 June 2016, askamathematician.com/2016/06/q-if-you-were-shrunk-to-microscopic-size-would-you-be-able-to-see-normally-would-you-be-able-to-see-microscopic-things/; "Imagicnation #02: What Would Happen if You Were Shrunk?," Steemit, 2017, steemit.com/science/@imagicnation/imagicnation-02-or-what-would-happen-if-you-were-shrunk; and Kurzgesagt, "Let's Travel to the Most Extreme Place in the Universe," YouTube video, 12:45, 4 October 2022, youtube.com/watch?v=FfWtIaDtfYk.

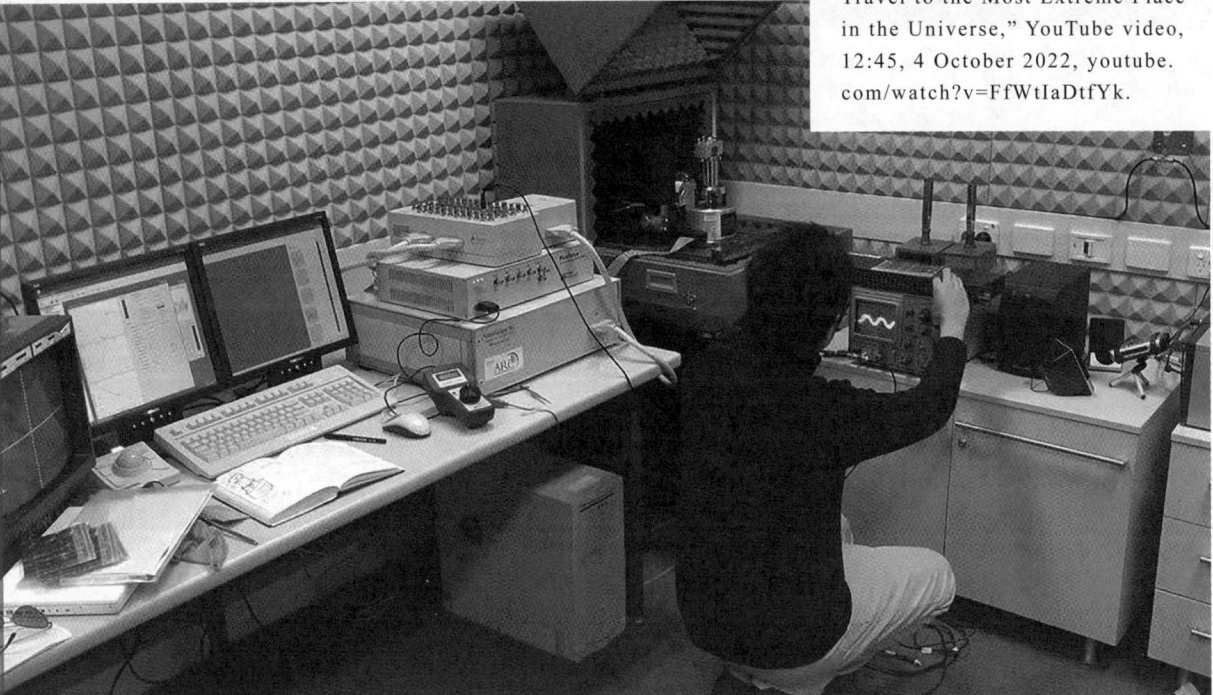

Joel Ong, *Nanovibrancy*, 2011. Site-specific sound installation and performance at John Curtin Gallery, Perth, Australia, 2011. Photograph courtesy the artist.

web. Only with electron and scanning-probe microscopes like the AFM can we see into the nanoscale realm. Of course, "seeing" might not be entirely appropriate here: Paradoxically, electron and scanning-probe microscopes operate in a state of blindness, employing a probing molecule to map the sample's topography by scanning its atoms row by row. The acquired data is then digitally processed into images through algorithms.

In *Nanovibrancy*, the AFM remains stationary, fixed on an artificial eardrum made from silk. The microscope is responsive to the eardrum's vibrations without causing it to move. Ong positioned the AFM's cantilever, a silicon nitride arm featuring a tip measuring approximately one-third of a millimeter wide, over the silk eardrum. As the silk subtly vibrated in response to its surroundings, the AFM made recordings, and the oscillations were scrutinized through laser reflections atop the cantilever. The voltage displacements were measured and then amplified and broadcast as multichannel sound into the gallery space, providing an auditory representation of the immediate acoustic environment as perceived by the eardrum.

Nanovibrancy reveals a universal facet of human perception, namely that our sensory input is a form of touch, and that all sensing therefore is a form of physical interaction. For instance, in vision, photons touch surfaces and engage with eye cells. Hearing results from air pressure waves physically impacting eardrums, while taste and smell involve molecular interactions with mouth and nose receptors. According to Ong, "At the microscopic level, sound entails the physical vibration of atoms and molecules, propagating waves through the air, and eventually impinging upon our eardrums. In *Nanovibrancy*, the AFM exposes this intricate soundscape by recording and amplifying nanoscale amplitudes of sound as the membrane assumes the role of listener."[19]

Everyday sounds would likely be experienced as extraordinarily intense in nano conditions. Sound frequencies would seem significantly higher, potentially in the ultrasonic or even hypersonic range. The body might even sense these sounds as intense vibrations or even structural movements within its tissues. Resonance effects and distortion might occur, and biological responses to sound, such as changes in cell behavior or tissue heating, might be more pronounced. While the exact quality of sound perception at such a scale remains speculative, Ong's work creates a powerful analog: "Part of me bringing it out of the science lab into a gallery space and recontextualizing the apparatus there was that the vibrations that were coming out of the performance were shared as a four-channel surround soundscape, and it was pretty loud. Obviously, you can't do that in a lab because the vibrations would skew your readings entirely."[20]

Nanovibrancy also recontextualizes the scientific experiment, taking it from the controlled environment of the lab to the dynamic space of a gallery. Moving millions of dollars' worth of scientific equipment into a gallery setting, Ong emphasizes the physical and experiential aspects of scientific endeavors. The volume of the transmission and its impact on the space and listeners poetically mimicked the volatile conditions of the nano realm and could be felt to the bone. The resonant frequencies and acoustic characteristics of the room intermingled with the transmissions, heightening the immersive experience and adding another layer of complexity to the sonic interplay. Thus, the performance's highly dynamic soundscape challenged the paradigm of scientific observation as a purely visual process. Static visual observation suggests permanence; it portrays its findings as objects rather than ongoing processes, as nouns instead of verbs. While images freeze time and allow us to meticulously examine samples that existed at a specific moment, the context and interconnectedness in which events unfold remain concealed. Before discovering acoustic ecology as a biology student at the University of Singapore, Ong recalls he would "go

19. Ong, "Nanovibrancy: An Auditory Exploration into Nano-scale Resonance," Vimeo video, 10:26, 28 July 2011, vimeo.com/27036778.

20. Ong, interview with the author, 24 August 2021.

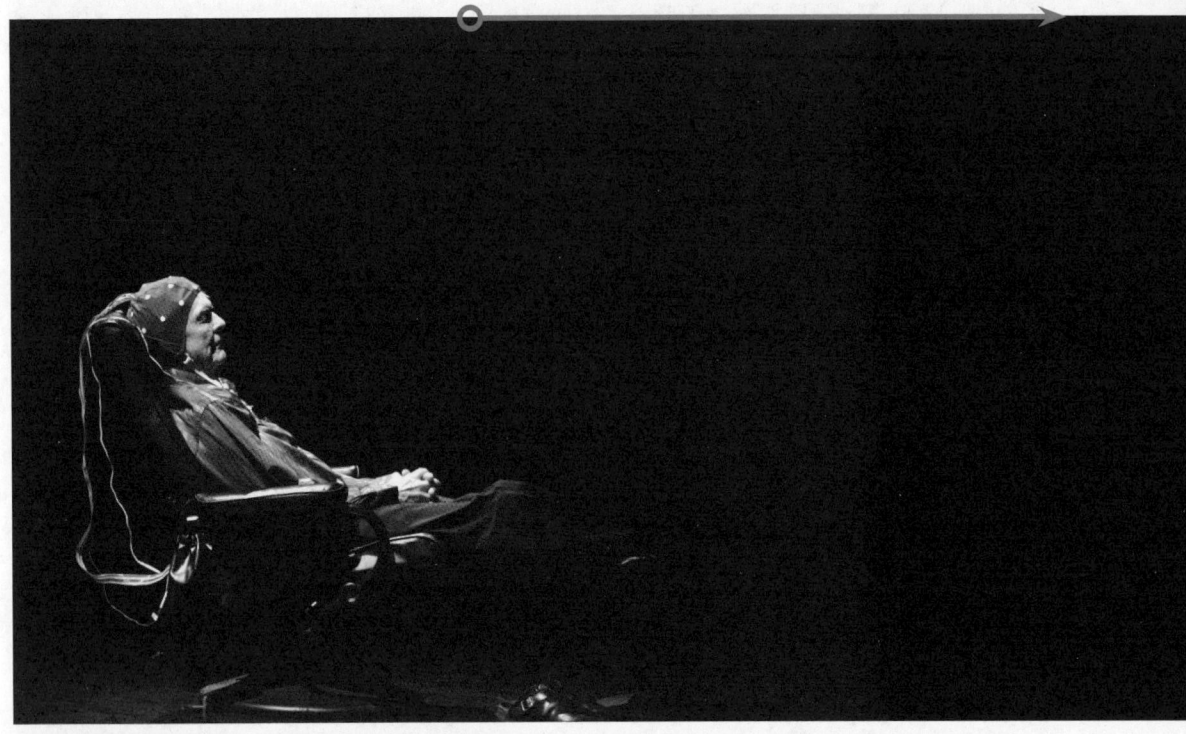

Juan Pampin, *Hemispheres*, 2016. Electroencephalogram performer Stuart Dempster with 3-D audio projection of brainwaves at Meany Hall for the Performing Arts, University of Virginia, 2017. Photograph courtesy the artist.

out into the environment and think about it quantitatively. Let's measure conditions, let's analyze intertidal zones. Up until that point, I never realized that this information could be shared through sound. Acoustic ecology became for me a way to go into the environment and understand it the same way a scientist would but through sound."[21]

Central to *Nanovibrancy* is an exploration of the complexities of human hearing, where the ear simultaneously functions as both the mechanism for and the content of the listening process. This arrangement resonates with Merleau-Ponty's concept of the self-touching hand, as the act of listening becomes an experience that is as introspective as it is outwardly engaging. Ong conceptualizes the notion of the ground of listening, membranic resonance, into a kaleidoscopic array of membranes on display and otherwise involved in the performance. The interplay between the eardrum membrane (the silk sample), the translating membrane (the speaker), and the receiving membrane (listeners' ears) interweaves perceptual models, creating a multidimensional topography of membranes sending and receiving. Furthermore, the

bodily experience of sound vibrations brings awareness to the corporeality of sound, metaphorically placing the listener into the alien nanoscape. Bridging the gap between scientific experiment and artistic expression, *Nanovibrancy* points at the changing nature of academic disciplines while prompting audiences to reflect on the interplay of sound, perception, scale, and space. As Ong summarizes, "Observing these small systems reminds us of our place in the larger environment and our role within a vast global system. The work not only offers an artistic perspective on nanoscale phenomena but also prompts us to consider our position in the grand tapestry of existence."[22]

21. Ibid.

22. Ong, in "Jack Straw New Media Gallery Interview: Joel Ong," Jack Straw Cultural Center website, jackstraw.org/artist/joel-ong/.

HEMISPHERES:
THE BRAIN IN THE SKY OF OUR MIND

Multimedia artist Chun Shao reclined on an Eames lounge chair in the corner of an old brownstone building at King Street Station in Seattle. On her head, she wore a red cap encircled with electrodes connected via cables to a mixing and control panel. Composer Juan Pampin instructed, "Imagine you're in the forest. Relax, listen to the birds." Shao followed Pampin's direction, listening attentively to the soundscape that was being projected three-dimensionally into the space. Pampin sat opposite Shao across the room, behind the control panel. Nearby, empty chairs awaited the audience that would soon witness their efforts, the performance *Hemispheres* (2016). Pampin then directed, "Now, try to predict the next sequence of sounds and then disengage from listening and calm your mind."

Shao's cognitive processes, as she attuned to different aspects of the soundscape, were recorded through the electrodes. Initially, Pampin recorded her alpha waves. These are the most dominant brainwaves, in the frequency range of eight to twelve hertz, and indicate a state of deep relaxation and passive attention, comparable to meditation. Pampin then recorded Shao's delta, theta, and beta waves. Theta and delta waves, characterized by frequencies ranging in the lower registers from 0.5 to eight hertz, signify states of extreme calmness and sleep, respectively, while beta waves, with frequencies ranging higher than alpha waves from twelve to thirty-five hertz, signify wakefulness and active outward-directed attention.

Pampin, who is a professor and chair of the Department of Digital Arts and Experimental Media at the University of Washington, employed simple click sounds and sine tones to acoustically represent Shao's different brainwave signals when their amplitudes exceed a specific threshold. Consequently, the interplay of signals from various brain cortices generated a staccato-like pattern of polyrhythmic pulses, which, while exhibiting a subtle and evolving periodicity, was not danceable. "This whole idea of rhythms to my surprise is reflected in scientific terminology," he explains. "So, when the performer is relaxed with eyes closed scientists refer to it as the *posteriorly dominant alpha rhythm*. The rhythms we detect are neither entirely random nor precise; they move around the median of a band and tend to have structure. Of course, it's hard to say it's a 3/4 against a 5/8, but they tend to have regularity."[23]

While Pampin uses minimal and naturalistic sounds to represent the spikes in brain activity, he remains deeply interested in a creative and intuitive approach, where brainwaves serve as a tool rather than a self-contained theme. This distinguishes *Hemispheres* from its grandparent, Alvin Lucier's *Music for Solo Performer* (1965), the first artwork to use brainwaves to generate music. Lucier's alpha signals, detected by two electrodes attached to his head, directed a series of electromechanical percussive instruments. Once Lucier relaxed enough to reach a state of steady wave flow, the work primarily explored non-intentionality and noninterference. "You get into a state that produces waves and then you just let it go. Some parts of hemispheres are very similar, but composition and artistic expression are the main protagonists of hemispheres." But Pampin, rather than simply observing brainwaves, chose to incorporate changes in their activity back into his composition.

A key feature of *Hemispheres* is its use of three-dimensional sound. Over the audience's seating area, a dome-shaped rigging truss suspended twenty loudspeakers, mirroring the electrodes encircling Shao's head. Pampin began the performance with simple sounds he described as "lab sounds," sine tones that occur singly or layered, modulating each other, as well as synthesized sounds from an analysis of a bandoneon, an accordion-like instrument developed in nineteenth-century Germany and predominantly found in Argentine tango music. While the piece unfolded, Pampin mixed in more complex waveforms, eventually incorporating field recordings of forests, wind, and water. Shao began shifting

23. Juan Pampin, interview with author, 17 May 2022. Subsequent quotes are from this interview.

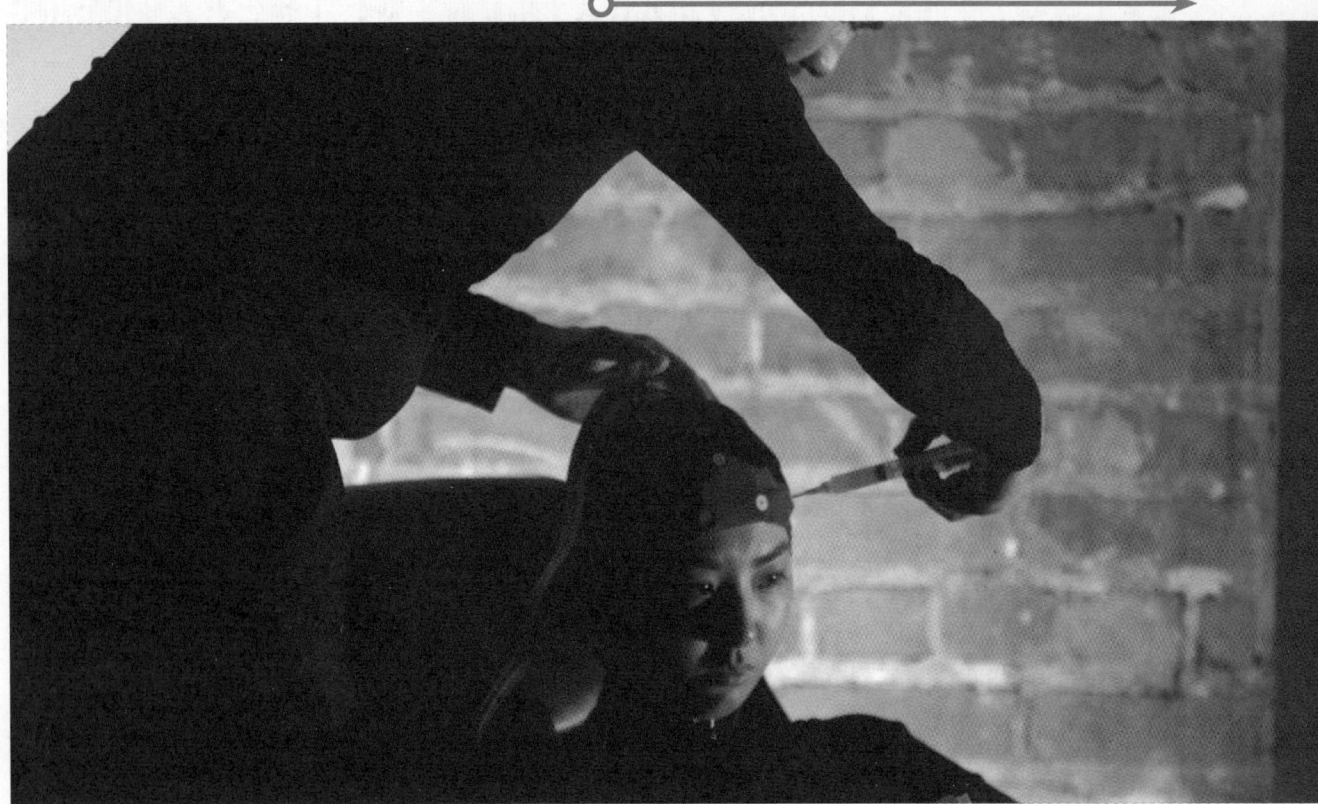

Juan Pampin, Hemispheres, 2016. Electroencephalogram performer Chun Shao with 3-D audio projection of brainwaves at King Street Station, Seattle, 2016. Photograph courtesy the artist.

her attention between the field recordings and the sounds produced by her own brainwaves, as the piece became sonically fuller, creating a feedback loop of action and reaction. Over time, Shao's visual cortex became more active, and the audience perceived an increase in sonic activity behind and above them. And when she was emotionally engaged, the sounds moved to the front and above, echoing the activity in Shao's brain regions. The result was an elaborate piece of sonic geography in which individual layers unfolded and interacted with one another, conceptually mirroring the hemispheres of the brain.

For a second performance of *Hemispheres*, trombonist Stuart Dempster took on the role of the performer. Dempster is cofounder, with composers Pauline Oliveros and Panaiotis, of the Deep Listening Band, a musical ensemble established in 1988. Their ensemble explored the concept of deep listening, an attentive awareness of sound and the environment, by creating immersive and improvisational sonic experiences that encouraged listeners to connect with their auditory

surroundings. Dempster is well accustomed to immersive states of deep relaxation as well as intense focus, and he was instantaneously able to access and modulate these states when he was connected to *Hemispheres'* electroencephalogram. As Pampin recalls, "The incoming signal was so strong, I had to use a compressor. But Stuart knew how far he could go. If you have practiced meditation or deep listening, you know how deeply you can delve into these states and how to enter and exit them. It's like controlling your brain in a specific way, and that's how this instrument works. He could have these subtle nuances that were quite delicate."

While *Hemispheres* might not explicitly reveal the performer's inner thoughts or emotions, it does showcase distinctive aspects of their personality, much like playing any other musical instrument would. The performer's unique traits have a significant impact on the work, shaping it in diverse and unexpected ways. "How does one define the piece? It is not a composition for a specific performer, and each interpreter would create a new version of the piece," says Pampin. "Hence, it could be entirely different. This aspect fascinates me in terms of what can happen with something that has certain parameters defining its form, but you don't really know how bumpy the road will be."

Hemispheres initiates its exploration with the brainwaves of the performer, positing these vibrations as the fundamental essence that constitutes our self-perception as conscious, self-aware entities engaged in the exploration of the world. Pampin's deliberate choice to project this universally resonant yet profoundly personal domain onto a celestial vault highlights a dual proposition: First, it confronts us with the externalization of what is presumed to be the source of interiority. Second, and herein lies the inherent paradox, the installation evokes the notion that our conscious experience is underpinned by a sense of infinity, a vast expanse open to all possibilities. This metaphoric juxtaposition not only challenges our understanding of consciousness and self-awareness but also invites a reevaluation of the interplay between the internal and the external, the finite volume of matter that our bodies inhabit and the infinite expanse of the universe which we are all able to experience within our bodies.

OLD SENSES, NEW SENSIBILITIES

We have yet to develop human senses that can detect our own brain activity, nanoscale sonic events, or electromagnetic waves, let alone see a black hole or feel a gravitational wave. But technology lets our minds venture into these realms, and we can make the forces and vibrations that surround us tangible through art and metaphor. Our ability to engage in creative analogy, artistic expression, poetic articulation, and non-disciplinary research collectively engenders new sensibilities. This emergence occurs not in isolation but through the amalgamation of these various facets. Artworks like *La Paracantora*, *Nanovibrancy*, and *Hemispheres* present intangible phenomena within a specific gestalt; the data collected by scientific instruments is but one constituent part; the body is another. Site, dimensions, appearance, social setting, and any other parameter deemed vital for the

work by the artist also have a part in how we come to feel and think about our relationship to the intangible.

Beyond illustration, these works transcend the typical purview of conventional scientific representation, which aims to objectively visualize data and often disregards connections to human experiences and entities not immediately related to that data. To hum along with the electronic "vocalizations" of a piece of technology, to be fully submerged in sonic magma, or to be reclining under a sky of acoustic brainwaves provides a vantage point no chart or table can offer. Furthermore, these works' real-time translation of data, whether through sonification or visualization, significantly deepens our understanding of the natural world by providing immediate sensory feedback. This instantaneous interpretation offers an intuitive comprehension of data and phenomena as they unfold, unveiling patterns, fluctuations, and relationships that might otherwise go unnoticed. It offers us the opportunity to relate to them in our own temporal dimension. The immersive experience enables us to discern subtle nuances and trends, thereby enriching our comprehension of the complex mechanisms within natural systems.

A theory of embodied experience, particularly as articulated in Clark and Chalmer's extended mind theory, elevates the role of sensualized data in our comprehension of the world and self. It implies the transcendence of our senses within the margins of our biological corporeality into the realms of technology and scientific data. This pseudo-sensorial expansion, in combination and contrasted with a recognition of our own sensory limitations, empowers us to perceive and interact with the world in ways hitherto inconceivable. Engendered by artworks, such interactions blur the line between individual selfhood and external milieu, creating a holistic, interconnected perspective on our human condition. They bring us into a heightened mindfulness of the world around us and a sense of connection with the vastness of reality, simultaneously encouraging a humbler and more open-minded approach to the unknown. <

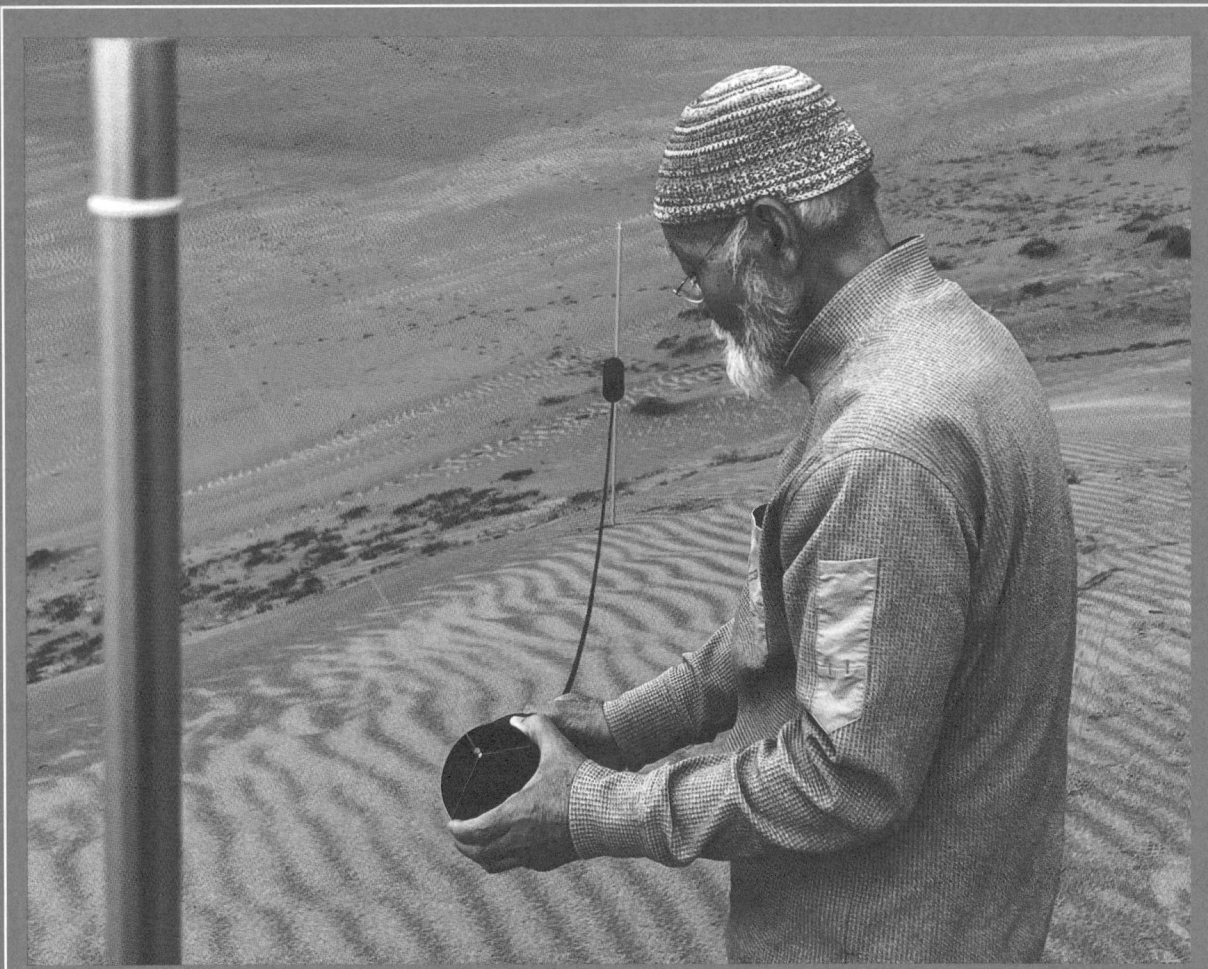

Akio Suzuki with the Analapos at the Tottori Sand Dunes, Tottori, Japan, 2023.
Photograph by Lawrence English.

LAWRENCE ENGLISH

in conversation with

October 27, 2023

translation by
ADAM SUTHERLAND

AKIO SUZUKI

LAWRENCE ENGLISH: How has wind as an energetic force operated in and influenced your work, especially work with your self-created instrument, the Analapos?

AKIO SUZUKI: I originally played the Analapos using my voice. The sound of my voice reverberated through the instrument. In 1981, I was invited to do a photoshoot for a magazine, and we traveled to the Hamaoka Sand Dunes. At some point during the shoot, I set up the Analapos, and I could hear a sound I hadn't heard before. The instrument started producing an intriguing sound, which I found completely fascinating. This was my first discovery of how the Analapos held a special relationship with wind.

LE: I can imagine it was quite fascinating to hear something so unexpected, especially after you'd been working with the instrument for so many years.

AS: It really captured me! In 1982, I returned to the Hamaoka Sand Dunes for an NHK film project aimed at introducing my work overseas. I was determined to re-create the sound I had heard before, it was so special to me, but I struggled to re-create what I'd previously heard. After sticking at it for a whole day, I learned it was the wind's physical interaction with the instrument that was responsible for the sound I'd heard. As the wind passed over the spring, it created a very special tone that resonated through the instrument. Even now, there's still a lot about it I find mysterious.

I had almost given up on the idea of being able to record the sound that day when suddenly I heard it once more. We were trying to record the sound, but there was a strong wind, which meant considerable noise was being picked up by the microphones. I distinctly remember the sound engineer burying these high-end pieces of sound equipment in the sand while recording.

LE: This event is interesting to me because, for the first time, you were able to be an audience for the Analapos, not its performer.

AS: Yes, and it was decades after I produced the Analapos that I first discovered it could be played by the wind, too. I felt a deep sense of joy encountering that Aeolian harp-type sound. It was the joy of being liberated from the performative and, in some sense, artificial way of making sounds I'd been experimenting with up to that point.

Since then, however, despite my best efforts, I've only managed to hear the wind playing the instrument a handful of times. The strength and direction of the wind as it caresses the long coil spring connecting the cylinders on either end is critical. It's difficult to find the right conditions for that to take place. The angle at which the instrument was suspended across the dune between the two rods, the weather, and so on, all affected the result. I feel as though I'll need to keep pursuing the perfect physical conditions that made the sound reverberate through the space as it did before.

LE: One of the main focuses of *Energy Fields: Vibrations of the Pacific* is considering how vibration works on and affects our bodies and our world. I know these ideas have been a part of your work for many decades, so I'm curious to understand the connections you feel vibration might have on your approach of "throwing" and "following," and on your practice more generally.

AS: My work has been guided by throwing and following, but that only reaps meaning when it is put into practice in nature. You can have ideas you think will go one way, only for reality to bring you an entirely different, unexpected answer or result. It was this realization that led to my Self-Study Events of the 1960s, through which I explored echoes. Echoes travel through the air to our ears via sound vibrations and create a connection to a place or an environment. To me, this explains the relationship of the body and the world—sound becomes the link between them.

LE: *Space in the Sun* [1988] reflects these ideas about our relationship to place quite strongly. I recall you once described how the energy of sound travels not just through the walls, but also around the surface of the planet—conceptually at least. How does energy and vibration operate in and on this piece?

AS: For *Space in the Sun*, the idea was to sit between a pair of walls for a whole day and simply experience the sounds of nature. For concentration purposes, I selected the northernmost point of the meridian line that determines Japan's standard time as the site for this practice. Constructing the piece took two years of hard labor; it was one of the largest projects I ever undertook. The performance was undertaken on the autumnal equinox.

The underlying concept was that of imagining virtual sound, as opposed to the idea of virtual images, returning as echoes along the 135th eastern meridian, following earth's longitude through Adelaide and back around the globe to the constructed walls, which were situated atop a mountain ridge 150 meters above sea level. It was a very important work for me.

LE: When I think of *Space in the Sun* and other pieces you've created, quite often I am thinking about the idea of reflection, of the sound's echo returning to you from some material surface you have thrown it out to. I am especially interested in the space and time between when a sound is created and when it is heard, and between when a sound is initially heard and then heard again reflecting off a surface or body. What has your practice taught you about how sound operates in these moments of in-between, when it is yet to arrive reflected to you?

AS: One might call this the space or pause or ma between throwing and following. Once a sound is created, the phenomenon that is its unfolding must be accepted and perceived with humility. It is a stance of waiting without acting egotistically. I believe this stance forms the essence of the *ma*.

LE: Could you expand on your idea of waiting without ego? I want to understand how place might influence where and when you undertake this process of throwing and following and then waiting. Does a place need to offer an invitation to throw?

AS: Initially, I engaged in practices based around throwing and following in anticipation of hearing an answer. But with my Oto-date work, which came later, I picked several throwing spots in public for the express purpose of listening and presented people with a way of listening that was both passive, in that they were just standing there, but also active at the same time. I took what was originally a personal, self-cultivating act and turned it into a piece of public art. At every Oto-date spot, there exists a time and space unique to the individual standing, waiting, there.

LE: How have other energetic forces and the ideas of vibration influenced and impacted your work? So many of your installations, for example, are concerned with kinetic energy. Can you speak to how your work recognized these concepts?

AS: I've always been interested in the unique vibrations that each material contains. I suppose you could say that, throughout my career, my approach to performance has been to draw out and listen to the nature of the materials I use. Many of my installation works are silent, but sound is always there, present.

There is a famous haiku: The ancient pond / A frog leaps in / The sound of the water. There are an infinite number of ways you could imagine this, depending on the shape of the pond, the time of day, the type of frog, the volume of the splash, and so on. I've always believed that it's possible to express sound without actually making any sound, and this has guided so much of the work that I have made across my life. <

by
VANESSA KWAN

Today, sound moves differently on the river than it has before. "It's climate change," Paul Josie says and gestures towards the shoreline. The evidence of erosion runs all along our route upriver. As we get back to the townsite, he shows us the clusters of willow trees that have sprung up quickly this season, spurred on by warmer summers. Thick swathes of them almost swallow the boat launch. Years ago, in winter, hunters could call up and down the river and be heard miles away. Voices and sounds would carry along the frozen shore and across

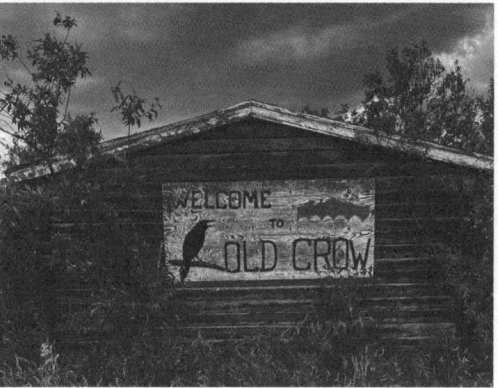

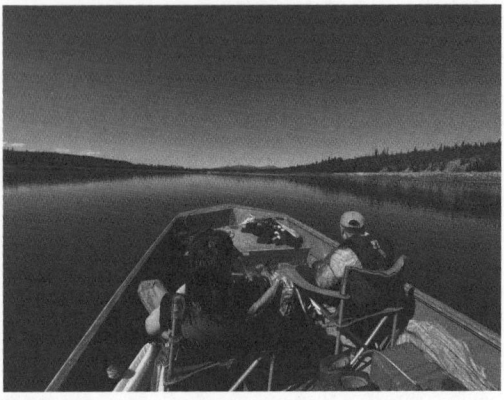

the snow for vast distances. Nowadays, the willow has changed how sounds carry, dampening the vibrations. In a warming climate, we are hearing things differently or not at all.

I'm in Old Crow, Yukon Territory, here to visit my friend and artist, Jeneen Frei Njootli, and their sixteen-month-old son, Nico. I'm travelling with Sarah Macaulay, who works with Jeneen through her gallery, and we're visiting from Vancouver, which is 135 miles south, on the west coast of Canada. Here, the Crow and the Porcupine Rivers converge, and together they flow nearly 300 miles to the Yukon River before finally emptying out into the Bering Sea.

Paul is giving us a tour of the river, taking us a few miles up the banks to a place called Cariboo Lookout. He is the grandson of Edith Josie, a well-known writer and journalist whose column, "Here Are the News," was syndicated in newspapers across the world and was a staple in the *Whitehorse Star* for over forty years. Edith's column beautifully detailed

the comings and goings of community members, hunters, tourists, and more through the area year-round. Caribou harvests were top of the news; life in the community is inextricably linked to the seasonal movements of the caribou, specifically a massive population called the Porcupine Herd that transits through the area and has since time immemorial.

Sarah and I are staying in Edith's cabin for the four days that we're here. Her writing table is still there, a green plywood table with a slight sideways tilt. We're on the territory of the Gwitchin, an Indigenous nation that spans Alaska, Yukon, and the Northwest Territories. Old Crow itself is a fly-in-only community of 250 people and the central community for the Vuntut Gwitchin.

We fly in on July first, a national holiday. There's games and food at the community center, and Sarah and I get to meet a bunch of people, play games, and eat sticky BBQ ribs and vanilla sheet cake. We win a couple of prizes in an egg toss and pudding-eating competition and lose miserably in the bubble-gum-blowing contest. It's eighty-eight degrees Fahrenheit, and at 1:00 a.m. the sun streams through the cabin window.

***

The previous year, in October 2022, I had written this in an email to Lawrence English and Robert Takahashi Novak about the project they are curating in Los Angeles for the Getty Foundation's 2024 Pacific Standard Time (PST) initiative:

Managed to talk to Jeneen ever so briefly today from Old Crow and

NOTES AND A

— 100

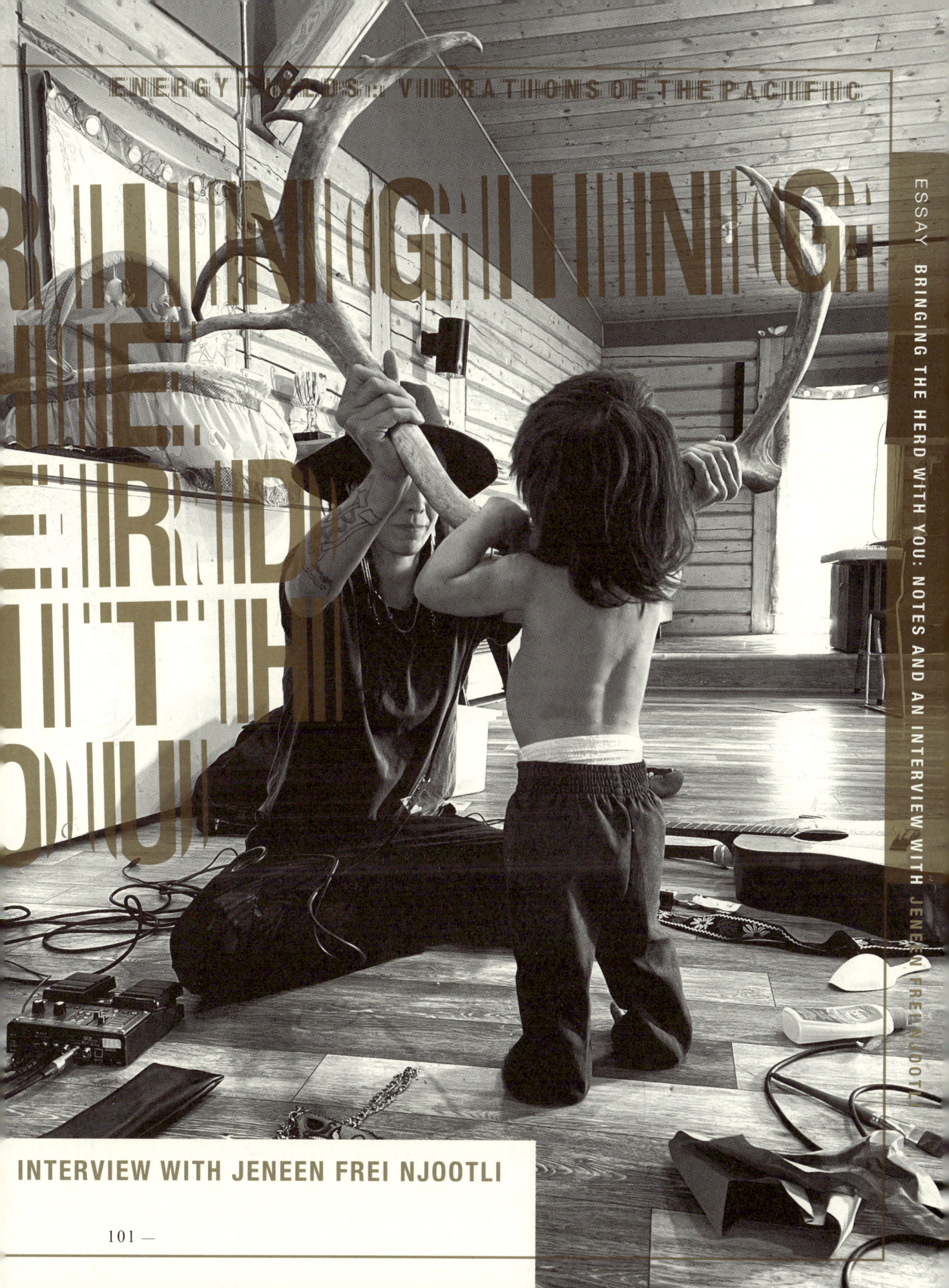

BRINGING THE HERD WITH YOU

INTERVIEW WITH JENEEN FREI NJOOTLI

told them about the project and that you are keen to include their work in the proposal for PST. They seemed really into the idea of being included in the exhibition proposal—and doing a performance too. They mentioned in particular a piece they want to do with a sound object from their territory: it's a moose shoulder bone with carved notches on it, and when it's played correctly it sounds "like a frickin' moose call, it's crazy" (their words lol). They are thinking about using machinery in the performance to carve the object, then play it for PST. Which would be so fucking incredible, if it could work out.[1]

***

Jeneen is an artist whose work I first encountered in 2015, in a gallery in downtown Vancouver. During that performance, they worked with piezo mics, looping pedals and amps to sonify a caribou antler and a guitar. As they drilled and cut it with power tools, scraped and dragged it, the antler became a micro-

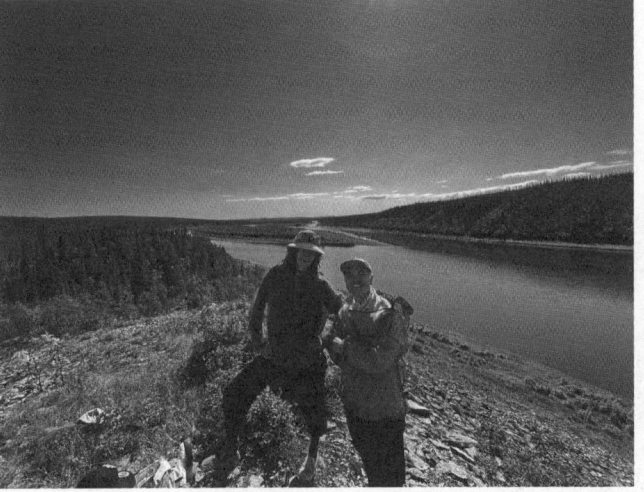

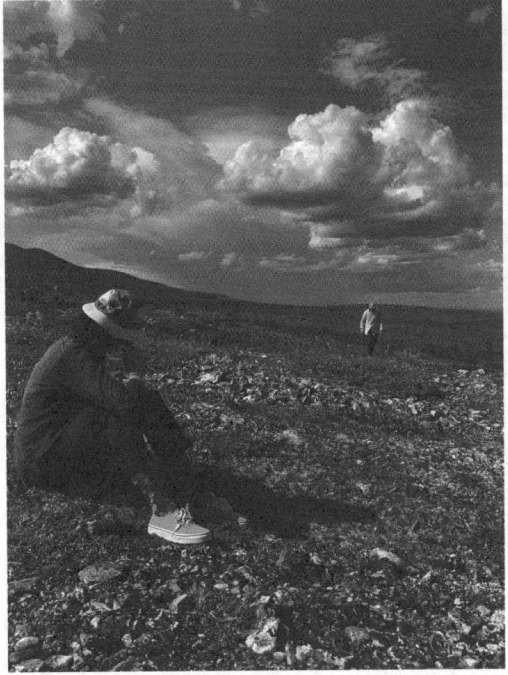

phone, and the transmission of sounds was both jarring and mesmerizing. During the performance, bone dust gathered over equipment and strings, and afterwards the gear remained on display, the fine powder a silent, gritty residue of the work. The leavings—organic, elemental, biological material—were unruly in the gallery, an airy cloud turned dusty coat. The way the antler produced sound as it transformed into particles created a disturbance that, for me at the time, redefined what could fill a gallery space. The connection to land and "somewhere else" was evident as a form of sonic longing, the dust a lasting impression of the performance, the sounds having dissipated.

A few years after that, Jeneen and I worked together on a performance in Seoul, South Korea. It was in many ways an uncomfortable experience—bringing Vancouver-based (at the time) artists across the Pacific to generate exchange. We stayed for two weeks in Seoul in the spring of 2018, and during that time Jeneen performed on a rooftop for a small audience and a vast cityscape. With pedals and contact mics, they sonified a hairpin made of a moose toe bone and beaded hide. Dragging the hairpin across the concrete roof and speaking song into the beadwork, Jeneen crouched on the roof, their back to the skyline. Working within the framework of

---

1. Email from author to Lawrence English and Robert Takahashi Novak, 11 October 2022.

a government-funded art project in a sprawling urban center made it difficult to reconcile the intimate content of the performance with the massive environment around it, however we might have tried. As one of the curators of the project, I felt out of my depth. I recall obsessing over aspects of the documentation, how to "capture" what was happening without objectifying the content, and how to translate both the nuance and the scale of the work into another language entirely. We had to move quickly to do the show, to install and print invites and rent equipment, and it felt like things were being lost. While we're in Old Crow, I interview Jeneen, and they talk about that performance:

That was an interesting one. Just thinking about expectation; to travel so far and to have such a short set or performance. [I]t was such a densely populated place [and] I am playing this... moose toe bone and this belonging that a relative of mine made for me and thinking about Indigenous material culture and how there is such an extractive relationship to our belongings. [F]or me, I [was] thinking about having bead work [sonically] fill space. I thought of how to make one small, beaded piece feel monumental. Or be too much, [such] that folks are forced to leave on a physical level... There's a certain level of discomfort that people can feel in my shows and...what does it mean to welcome that? And, also, if people [have] different thresholds for discomfort... when is discomfort productive?[2]

    Jeneen and I talk further about the choice to leave a performance. They came up through experimental noise communities in Vancouver in the early 2010s, where the polarizing nature of sound became evident. Sound works and noise shows affect people differently, and most often in a very embodied way: some folks fall asleep, and others are driven to anxiety, anger, or irritation. Interest in these different bodily reactions to vibration drew Je-

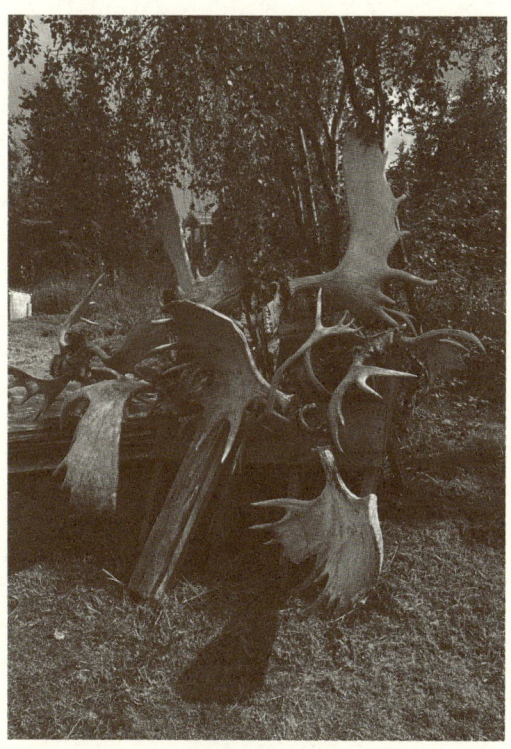

neen to sound work, and some basic principles around the impacts and opportunities of the medium were laid in early on: Noise shows compel a bodily reaction, leaving or staying is a choice, and you're welcome to do either. As they observe, "People leaving isn't a bad thing, and that's part of sovereignty and inviting people to have the right to be human, to be in their body, to be present with themselves. And what does support look like? Sometimes support can also look like leaving." But for those whose ears desire the sound, it's a different experience:

[F]illing a space with bead work feels political for me, [as does] taking up space with antler. I don't know if I want to use words like "disruption" or "challenging" or "pushing back," but those things are happening. And one time, I performed at a potlatch that Beau Dick had at [the University of British Columbia], and [Tsimshian dancer, choreographer, curator, and activist] Mique'l Dangeli was there, and I shared my sonified caribou antler and also had an electric guitar and some distortion pedals played through a guitar amp. And afterwards, she said that she could hear the land. Like she felt like she was transported to the

2. Jeneen Frei Njootli, interview with the author, Old Crow, Yukon Territory, July 2023. Subsequent quotes are also from this interview, unless otherwise noted.

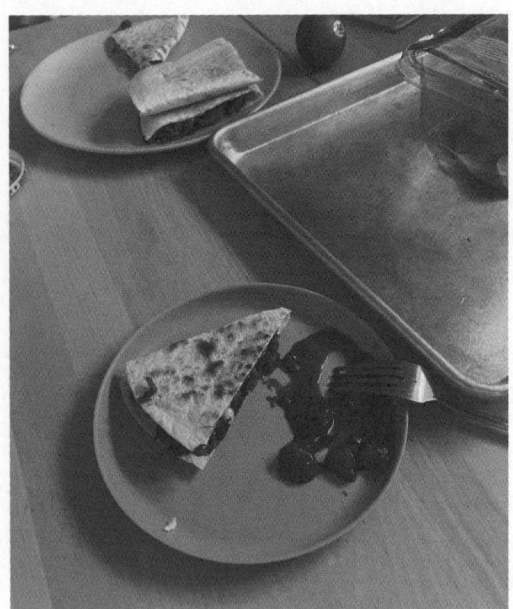

land up here. And in some ways, how could you not be when you're hearing sound through an antler?

The reception of the sound matters. Who is listening, and how they are listening, matters. Dylan Robinson articulates "critical listening positionalities," a "continuum of listening practices that include subtle and significant gradations of normativity."[3] This normativity sits in relation to what Robinson terms "hungry listening," a sensory paradigm defined by early moments of contact and the extractive insatiable desire for Indigenous lands, resources, and utterances. This hunger persists and is upheld structurally, but it also resides within each of us as subjects in how we listen, communicate, and create meaning. According to Robinson, "As part of our listening positionality, we each carry listening privilege, listening biases, and listening ability that are never wholly positive or negative; by becoming aware of normative listening habits and abilities, we are better able to listen otherwise."[4]

To practice resurgent or decolonial forms of listening, one's relational position as listener and *listened-to* matters deeply. This is where Jeneen's work asks questions: Who or what is being addressed? Sound is vibration, transmitted and received, but it's also much more than that—it's a reference and an ad-

dress to things both present and longed for. In Jeneen's practice, sonification stands in for a bigger conversation, one that expands across space, time, and decolonial positionalities.

In *Noopiming: The Cure for White Ladies*, Leanne Betasamosake Simpson, a Michi Saagig Nishnaabeg writer, scholar, and musician, imagines their supernatural protagonist, Adik, a caribou from a long-lost herd roaming a contemporary urban-settler world, witnessing one of Jeneen's performances. Seeking forms of connection, Adik travels their ancestral territory—the region known widely as Toronto and its surrounding suburbs—looking for traces and kin that exist amid the dominant knitwork of settler-built environments, social spaces, language, and worldview:

Adik stops into Artspace in Nogojiwanong on their way back from Kinomagewapkong to see Jeneen Frei Njootli's exhibition… [T]he best part was the opening, because Frei Njootli had invited the Porcupine caribou herd to attend with [them]. And wow wow wow. Here Adik was, in their own territory, dancing to Frei Njootli's sounds with the only living relatives they had ever known, while all the humans stood around acting impressed or dumbfounded or whatever they did when they attended art shows not really made for them at all.[5]

A moment of kinship across time and settler geographies is called in by Jeneen's performance, for Adik and others. Simpson points to something that can't be identified by hungry ears: that these vibrations don't care to be received in the way we assume sound or an artwork should be. It broadcasts with no fixed destination, but it is richly received

3. Dylan Robinson, *Hungry Listening: Resonant Theory of Indigenous Sound Studies* (Minneapolis: University of Minnesota Press, 2020), 3.

4. Ibid., 11.

5. Leanne Betasamosake Simpson, *Noopiming: The Cure for White Ladies* (Toronto: House of Anansi Press, 2020), 159.

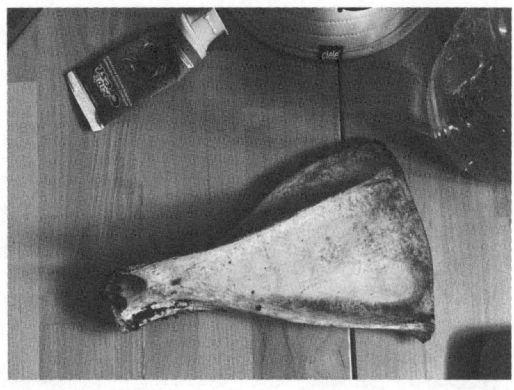

in spaces cracked open and divested of settler-colonial values; Jeneen's practice is held more fully with the knowledge those places afford.

The work recuperates what has been cleaved away. The wider conversation embedded in Jeneen's work is one that understands obstruction, extraction, and violence but doesn't actually want to look in those directions, because why? It isn't speaking to those things or us (and here I mean settlers who stand around at art shows). It's looking for a reception that is fulfilled by a connection to land, to place, to the caribou themselves.

***

I'm really grateful that our first introduction to community in Old Crow comes through game playing and snacks. Chips, Pepsi, and cake are things in common, things best shared.

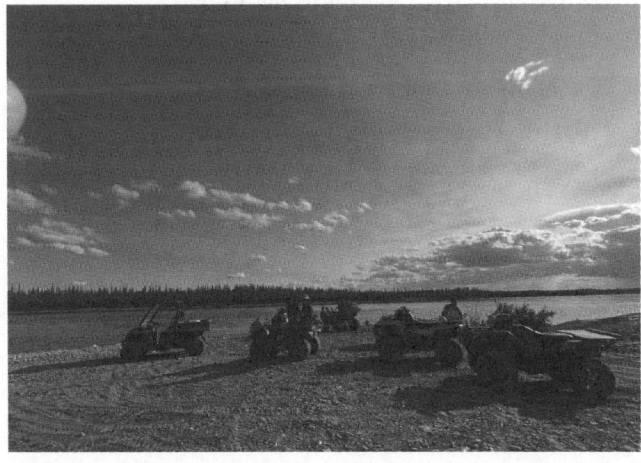

6. This term refers to a Ski-Doo, a snowmobile that is used widely for transport and recreation in Canada.

During the pudding-eating contest, I'm blindfolded, and an elder named Rosie shovels vanilla pudding into my mouth at an incredible speed (we win, and it's honestly one of the proudest moments in my lifetime). The pudding is snack-pack sized and sweet as hell. I haven't had it since I was a tiny kid. Perhaps taste is a bit like sound; the body remembers and desires it, even if it's been a long, long time.

Jeneen has meat defrosting on the stove. They haven't been able to harvest or find a moose shoulder, so they've opted to use a caribou shoulder for the performance in Los Angeles. Just before we head out to the community center, they put it in the oven for a low and slow session. By the time we get back, the meat has pulled off the bone and the house is filled with the smell of cooked caribou. They pull the meat into strips, offering up the connective bits where collagen and tendon attach to the joint. It's smooth and sticky and reminds me of the best, most unctuous parts of oxtail. We eat it over the next two days as quesadillas and breakfast tacos, adding Kraft singles and salsa and some chives that we harvested with Paul on the banks of the river. It's fucking delicious.

The shoulder bone itself is lighter than I expect it to be—the ridge of thick bone along the top, where the big muscle attaches, gives way to a winglike slimness before thickening again. The dense stub of the joint is still ragged with meat muck that hardens like leather as it dries. The shoulder is slightly rough to the touch, and when you activate it, it produces a deep and carrying tone—like if you tap the base of your throat right in between the bones of your clavicle, that feeling is the sound it makes. When it's dry, Jeneen uses a Dremel to clean off the last bits of meat and cartilage. We strategize about how to transport it to Los Angeles for the performance. Sarah laughs and says, "Give it to the white lady; I'll tell security it's a beauty tool." I think about all the people and things that are needed to get this piece of caribou from here to there.

***

Jeneen writes to me later, after we get home to Vancouver:

I wish u got to hear my nephew [who] you didn't meet call geese and imitate skidoo[6] sounds. It's incredible. That calling animals is something we've always done. The animals here used to understand Gwich'in… maybe they still do. There's a story about a woman at Rampart House who spoke to her dog in Gwich'in. That woman is my grandma. Gee I wish to find the story now for you. But thinking about the sounds our land misses. The sounds the belongings waiting in archives are waiting ready to hear n have been missing.

 Me n my dad did a dog whip project. They are old sound tools to communicate with dog teams. We went to [museums in] Ottawa and [Whitehorse] to visit w them. The old dog whips. I put on a workshop at the Heritage Centre here n we made 3 new ones based on research my dad did. It takes one generation for knowledge or a traditional tool to leave a community.[7]

 Words are a rhythm, like falling down stairs. Their meaning helps express an experience, and then they drop away when translation fails.

 The sounds the land misses. The taste of caribou. A dog whip.

***

Intergenerational antler jam: On our second day in Old Crow, Jeneen does a sound workshop for youth. It's stinking hot outside, and inside the youth center we bake pizzas and put out cheezies. Jeneen hauls their guitar, some cymbals, an antler, some beadwork, mics, loop pedals, and an amp over from their place. Everything is set up on the floor and a few kids show up from the community center. They jam for a while, and everyone wants to yell into a microphone and twist the dials on the amp. Over time, it gets gentler, and they start getting a bit tender, dragging the contact mic over things and pressing it against their throats and chests while they hum and whisper.

 "The beads are a microphone," says

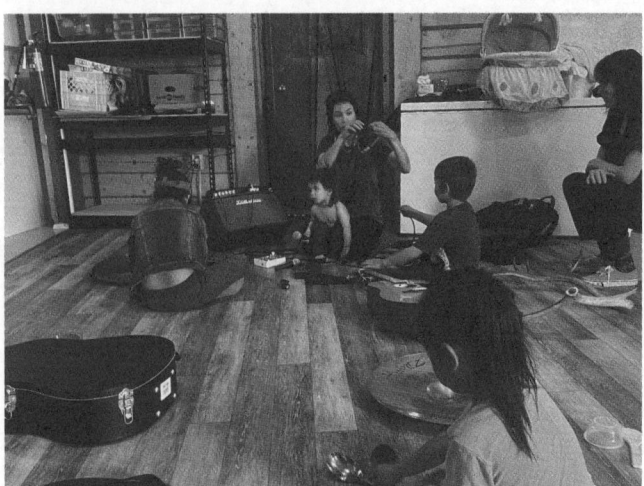

Jeneen, using a beaded belonging to make sounds and frictions. I'm not sure if the youth catch that, but Kylie, who is four, spends time drawing the mics across cymbals and starts quietly vibing with the loop and distortion pedals. Nico, by far the youngest, is fully engaged, and at some point, there's a full kid jam of "head and shoulders knees and toes." Nico and Jeneen sonify a massive caribou antler together, each pressing their mouths to it and smiling. We break for pizza, a few kids come and go, and by the end it's a mess of wires, juice boxes, bone, beadwork, and squashed strawberries squeaking under our sneakers.

I love working with youth, I love working with… Gwitchin youth in Old Crow. It helps bring a different confidence… when you're working with things… outside of the written or spoken language. Get[ting] your footing in that and feel[ing] confident is when you start to feel an understanding… Like Nico's growing up knowing how to make sounds with contact microphones, with his body. Anything that he finds can become a sound tool, and I think that it nurtures an innate knowing. And because it's something that's outside of traditional music notation or a set of instructions, it can encourage people to trust themselves. [I]t can nurture a trust, but it's also so vulnerable to lean into that unknown space of uttering.

***

7. Frei Njootli, Instagram message to author, 16 July 2023.

Is listening different up here? When we're doing the interview, Nico pulls the modem out of the wall. On the recording, you can hear Sarah and me scream, "Not the internet!" right as we're talking about the sound of caribou dew claws. The interview recording is garbled from then on, our conversation coming in and out of the muffled sounds of making snacks, feeding Nico, standing up, reshuffling our bodies, echoing whoops of toddler joy. The recording is a vibe, an atmospheric jam.

When Nico was born, Jeneen's relative, Debra-Leigh, made him a caribou rattle out of dew claws. It's a sound object that, like the moose shoulder bone, is used by hunters to call caribou during the hunt. When it's activated, the dew claws click together and replicate the sound of the herd walking. The animals approach, drawn by the sound of their own movement. Jeneen reflects, "[W]e've always had a relationship with sound and worked with parts of the animal to create sounds... to continue our relationships with them. [T]his is about continuing a relationship to the caribou... we listened for each other."

Pause. I'm realizing something as they say this. We privilege the vibrational choreography of language, the shared understanding of this word and that concept, of sound to meaning. But this is a different thing. This is a vibrational kinship, a bone-deep connection whose sonification allows for so much more than communication—it constitutes the survival of a wider body: of human, of animal, of land, and all. Jeneen uses vibration as a connective tissue; vibration-made muscle. What is shared in this system is tender, tough, and untenable within a settler colonial framework. It has always existed outside the bounds of that knowledge, and always will. As a practice of Indigenous resurgence, it pushes to inhabit the concepts that Dylan Robinson theorizes and the states that Adik longs for: an embodied and fully realized cultural expression where connection to one another and the land is not reclaimed but is as it always has been: innate. Where the impositions of settler-colonial logic and the violence of separation (from language, from culture, from oneself and the land) have neither place nor primacy. In my pausing, I understand that there are

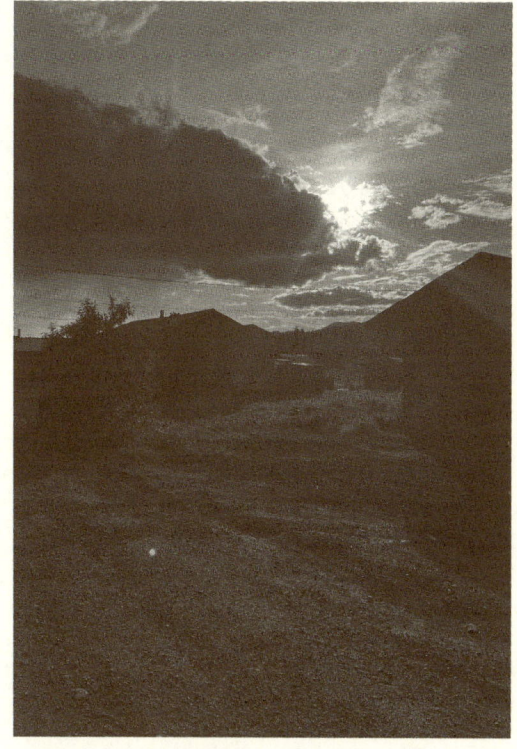

whole states of being held in this vibrational address; worlds expand in the practice.

I'm thinking, too, [about] doing this interview in my homelands... I always found being home really humbling; [despite] any success or achievement that I had in an art or in my career outside of my homelands... I would come home and then would be so bad at something. It was so important to not be strong enough to full-start a skidoo. I have some of this knowledge on my periphery or am in [the] beginning stages... but there's just so much to learn, still. The late Georgie Moses shared with me to never say that you're good at your culture, always be practicing. Then there's always more to learn and share.

[To Nico] What are you doing?

At its height, the Porcupine caribou herd numbers in the hundreds of thousands, and I try to imagine (and fall short) what it must be like to hear the clicking of thousands of dew claws crossing the frozen river. What would it be like to be drawn along by the sound of your kin, moving? So yeah, listening is different.

***

Back when we were in Seoul, one of the pieces Jeneen made for exhibition was a pair of beaded earplugs, referencing an innovation their dad, Stan Sr., made for riding his Ski-Doo. He would string his store-bought earplugs to a lanyard that was attached to the back of his baseball cap so he wouldn't lose them while moving at high speeds. There is swagger to Jeneen's work—a luxury interpretation, with gold-plated beads and high-value custom packaging. But the object, like its sound-based counterparts, hangs in an intimate balance of belonging. Jeneen invokes the specificities and adaptations of living fully intertwined; each is a communication of self to the herd, a conduit for being in conversation with the land.

Transposed into the urban centers of artworld dissemination and reception, their practice has other work to do, too. Using terms that fit nicely in my mouth and on my keyboard, I might call that word "resistance." But I've come to hate that word, even as I try to find ways to embody it. Jeneen's practice engages with what their land and people and belongings miss: the sounds and senses that have been forcibly taken away (and that now, in a climate crisis, are being inexorably

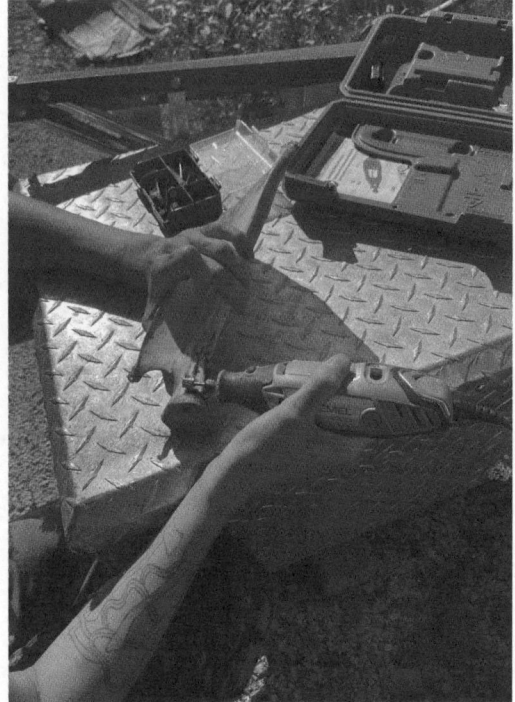

changed). This feels like much more than resistance, it feels like the hum at the base of a clavicle, the catch of a Dremel against bone, the place where a mouth touches a microphone, and the unknotting and re-knotting of a thin strand of hide. I won't presume to put a word to what this is.

Looking at it from another angle, the work upholds the offer of choice—whether you want to leave or stay.

***

I left. I went to Vancouver, and then to Los Angeles and back again. I took a stone from the spot where the Crow and the Porcupine Rivers meet (the river took Jeneen's sunglasses in exchange; I might still owe a debt.). The stone is red at the base, fits in the palm, is pockmarked and uneven on one side, with a thick band of cream-colored sedimentary deposit on top.

It looks like a piece of fatty meat cut quickly from the bone. <

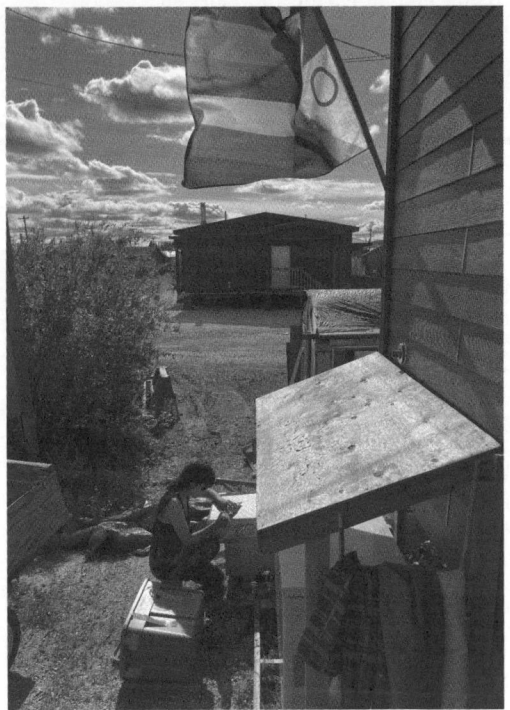

Photographs from the author's Summer 2023 visit to Old Crow, Yukon Territory, appear courtesy of the author, Jeneen Frei Njootli and Sarah Macaulay.

ANNEA LOCKWOOD

New Zealand–born American composer Annea Lockwood brings vibrant energy, ceaseless curiosity, and a profound sense of openness to her practice as a sound artist and composer. Her lifelong fascination with the visceral effects of sound in our environments and on our bodies—the way different sounds unfold and their myriad life spans—has served as a focal point for her recordings, performances, and multimedia installations.

For *Energy Fields: Vibrations of The Pacific*, Lockwood presents *World Rhythms* (1975), in partnership with Zebulon, and *Wild Energy* (2014) in partnership with the Los Angeles Arboretum. Both are sound works that draw on tones and vibrations gleaned from nature and the cosmos. *World Rhythms* is a nine-channel audio performance composed of field recordings of volcanoes, earthquakes, radio waves, geysers, and tree frogs, among other sounds. The multichannel *Wild Energy* was developed in collaboration with sound engineer Bob Bielecki, whose work creating custom solutions for recording and amplifying sound was central to shaping the piece. To create *Wild Energy*, Lockwood manipulated scientific recordings of phenomena including radio waves, solar flares, hydrothermal vents, and bat and whale echolocation to bring them into the human audio range. **<**

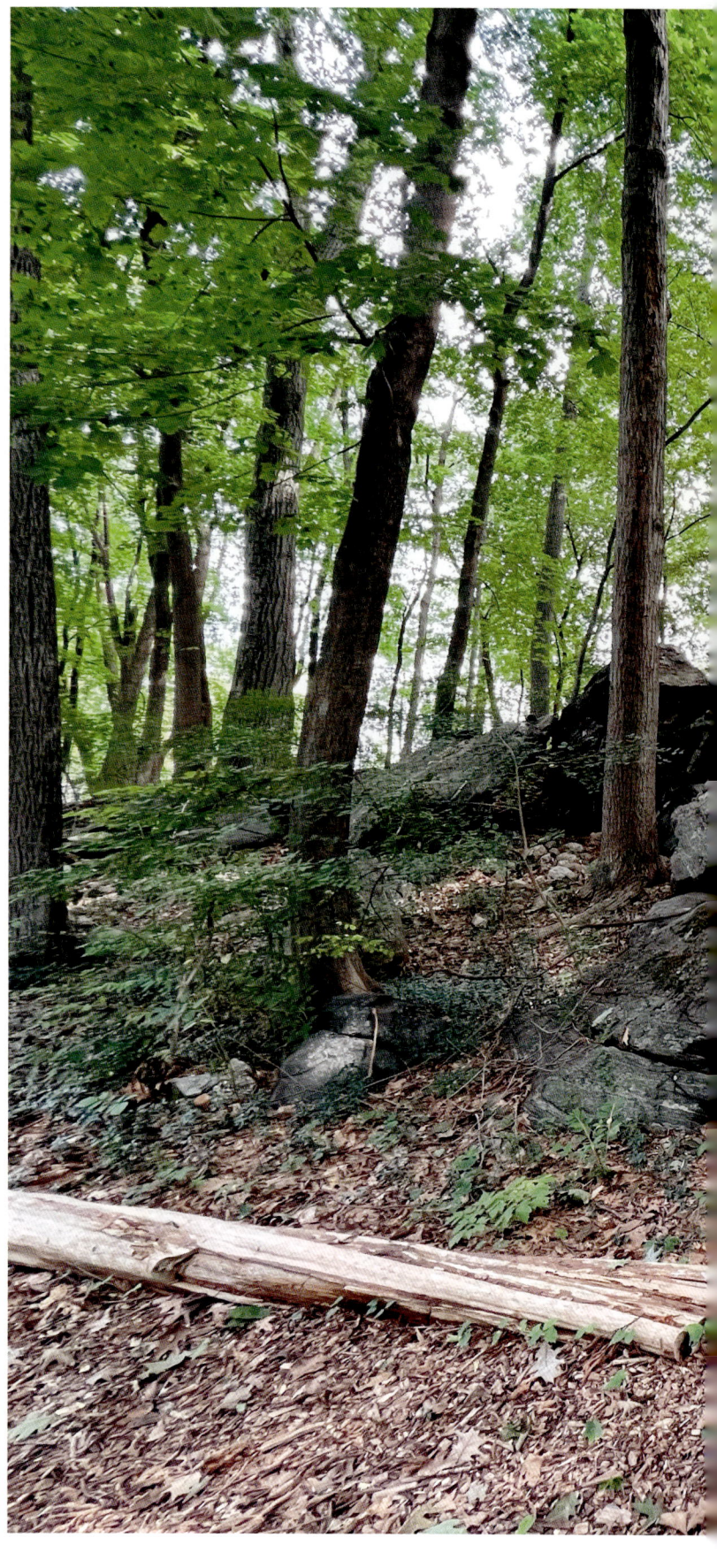

Annea Lockwood, *Wild Energy,* Caramoor, New York, 2023. Photograph by Lawrence English.

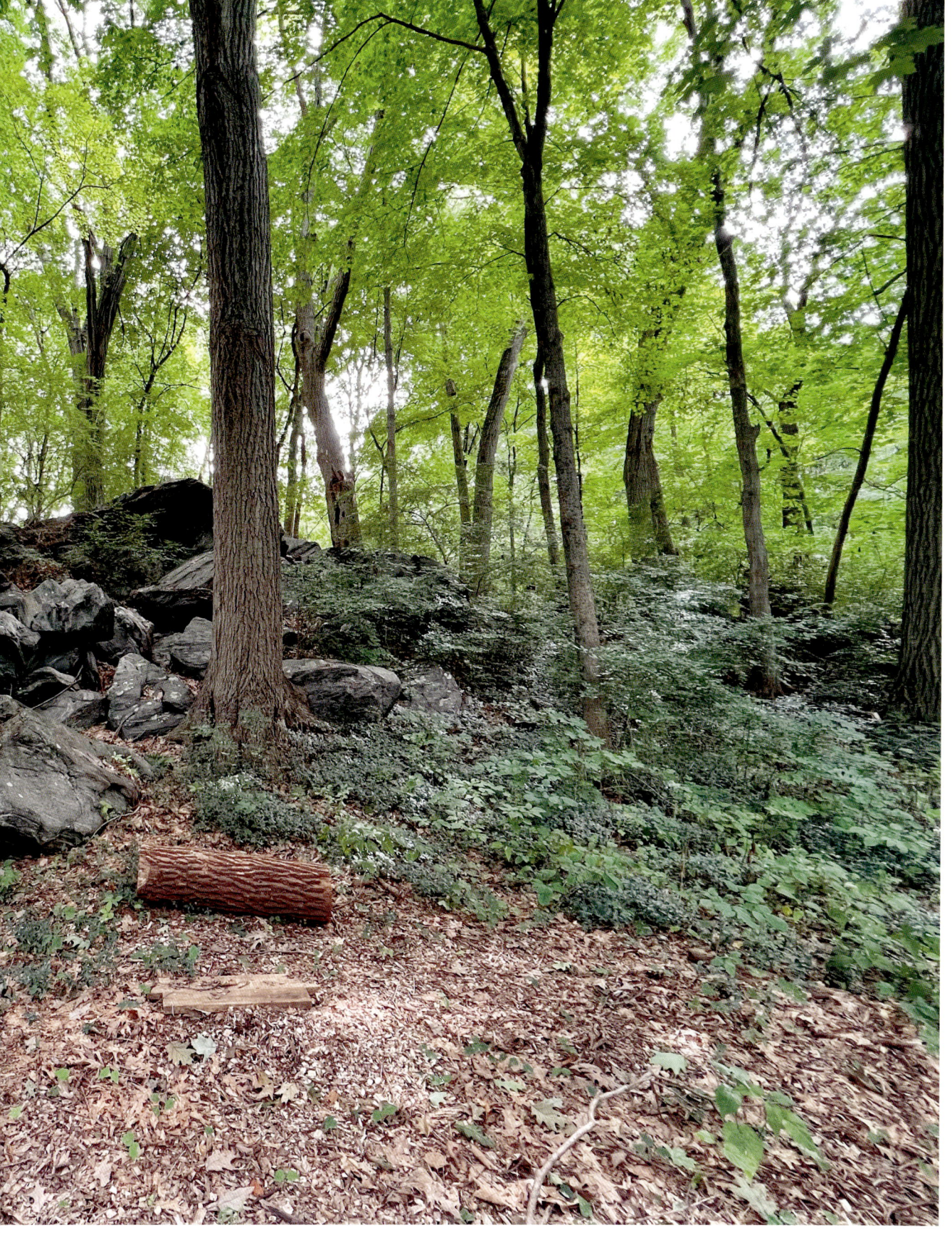

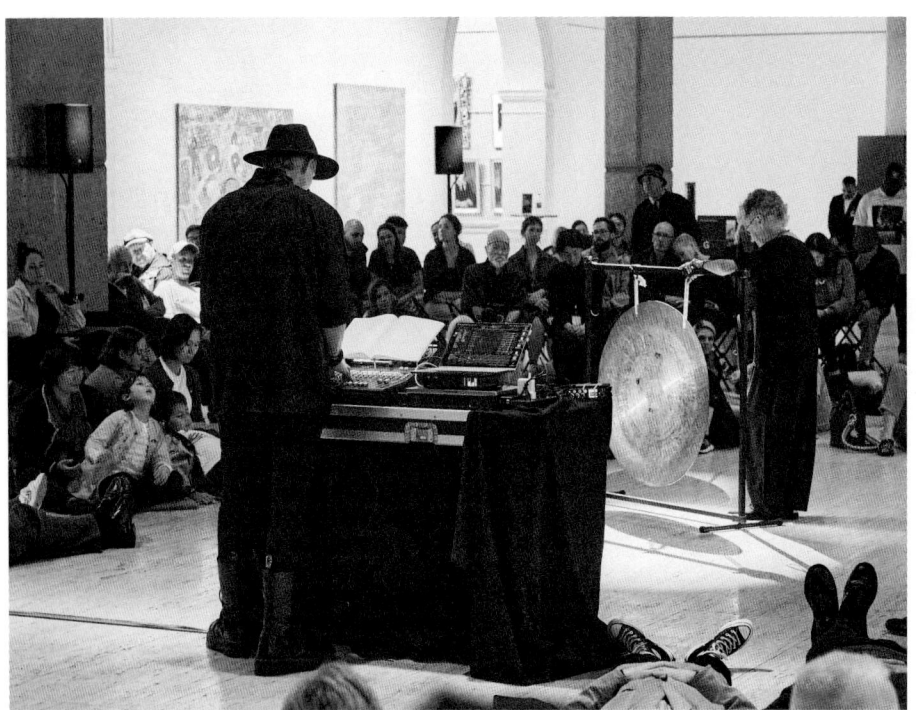

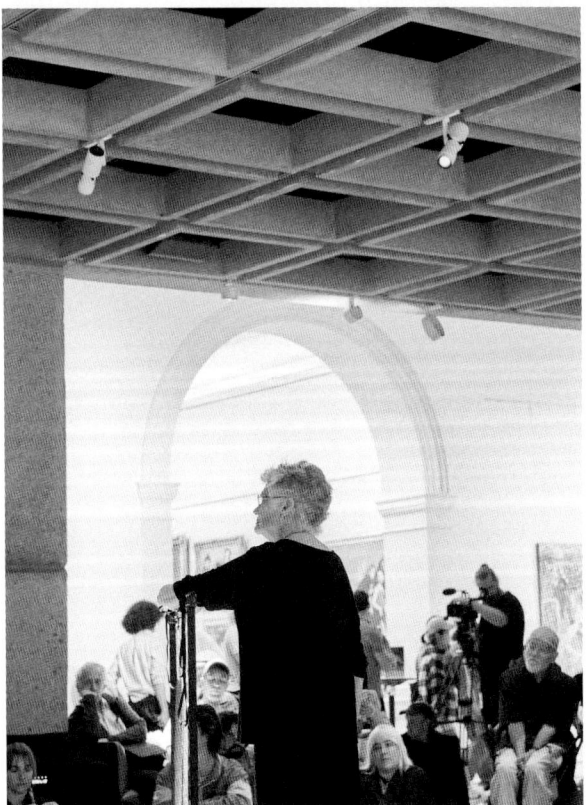

Annea Lockwood, *World Rhythms*, 1975.
Performance at the Art Gallery of New South
Wales, Sydney, 2023. Photographs by
T. Pakioufakis.

BETHAN KELLOUGH

Hailing from Scotland, Bethan Kellough is an artist and composer based in Los Angeles who uses string instruments, electronics, and field recordings to create immersive compositions and sound installations. Kellough employs a variety of recording techniques, including ambisonic microphones, hydrophones, and electromagnetic and VLF (very low frequency) receivers, to capture the vibrations and spatial nuances of the natural world. She is particularly fascinated by the sounds of geological processes, and she draws on these delicate and powerful vibrations to create works that are both intimate and expansive.

For *Energy Fields: Vibrations of the Pacific*, Kellough has been commissioned to compose and perform a new work, *Still the Fragments Move* (2024), presented in partnership with the Center for the Art of Performance at the University of California, Los Angeles. *Still the Fragments Move* draws on ambisonic and hydrophone recordings of the Pacific Ocean to explore the vibrational characteristics of ocean waves, from the play of wind on their surface to their encounters with rocks, pebbles, and sand at the shore. <

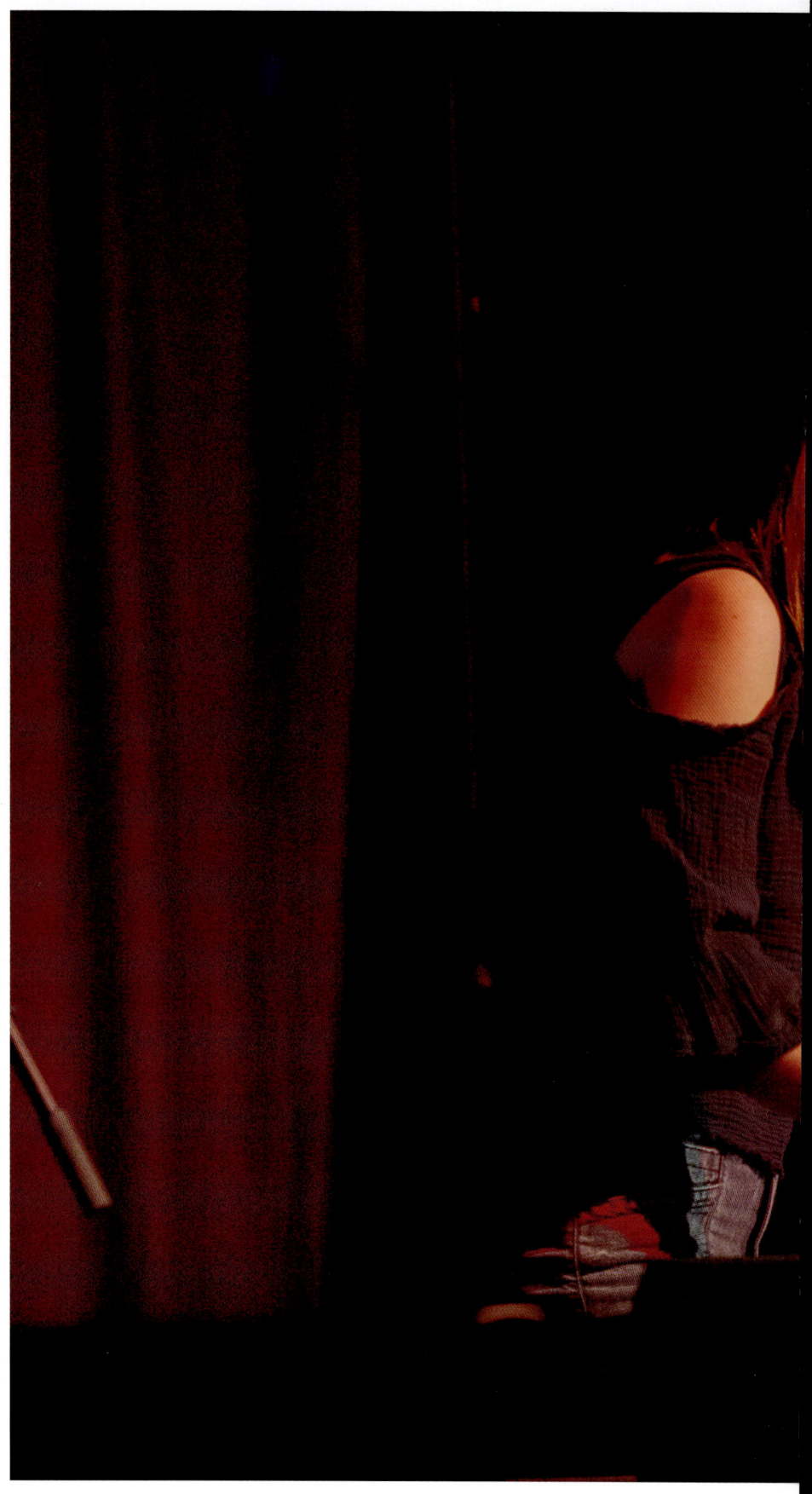

Bethan Kellough performing
at the Fulcrum Arts Festival,
Zebulon, Los Angeles, 2022.
Photograph courtesy of
Fulcrum Arts.

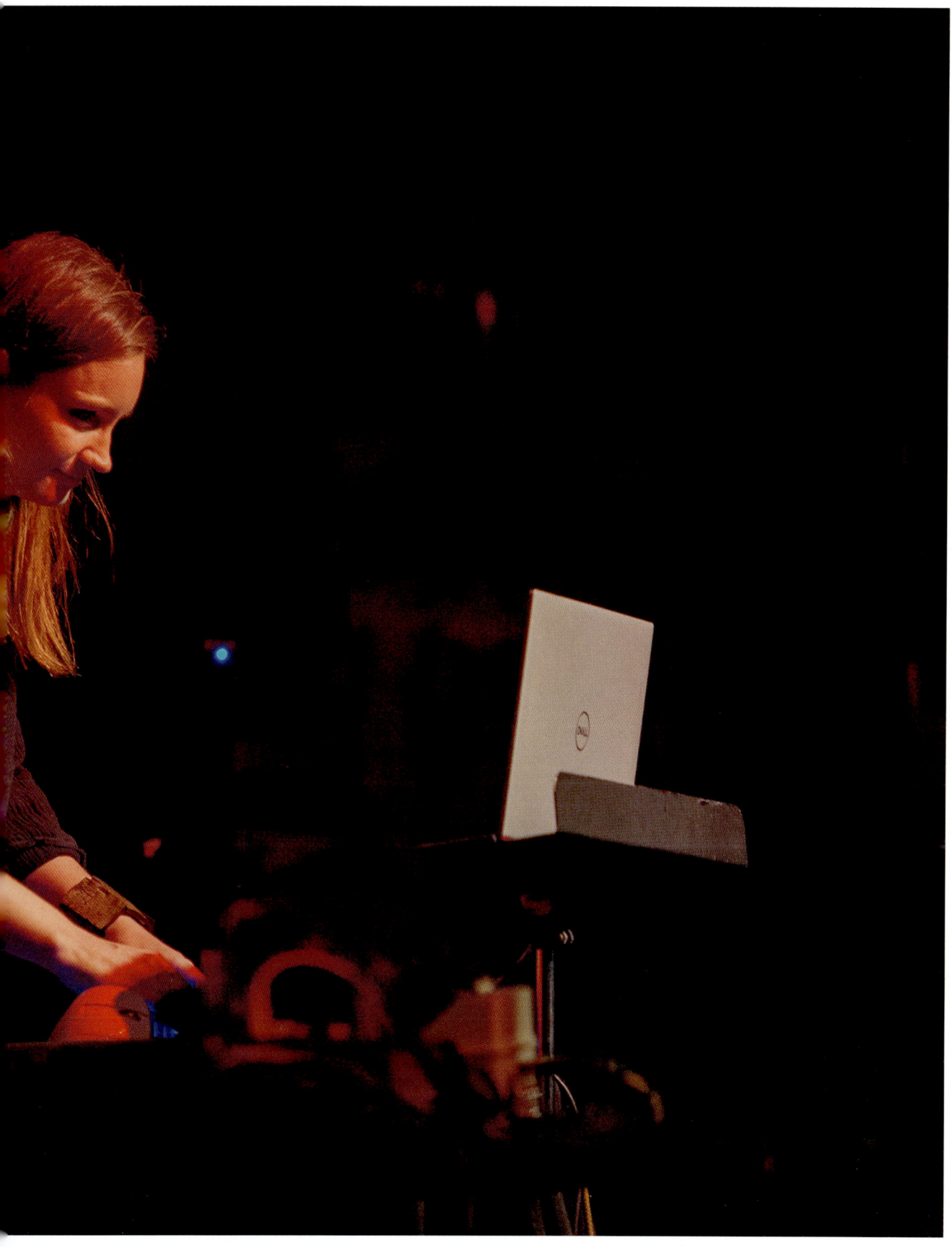

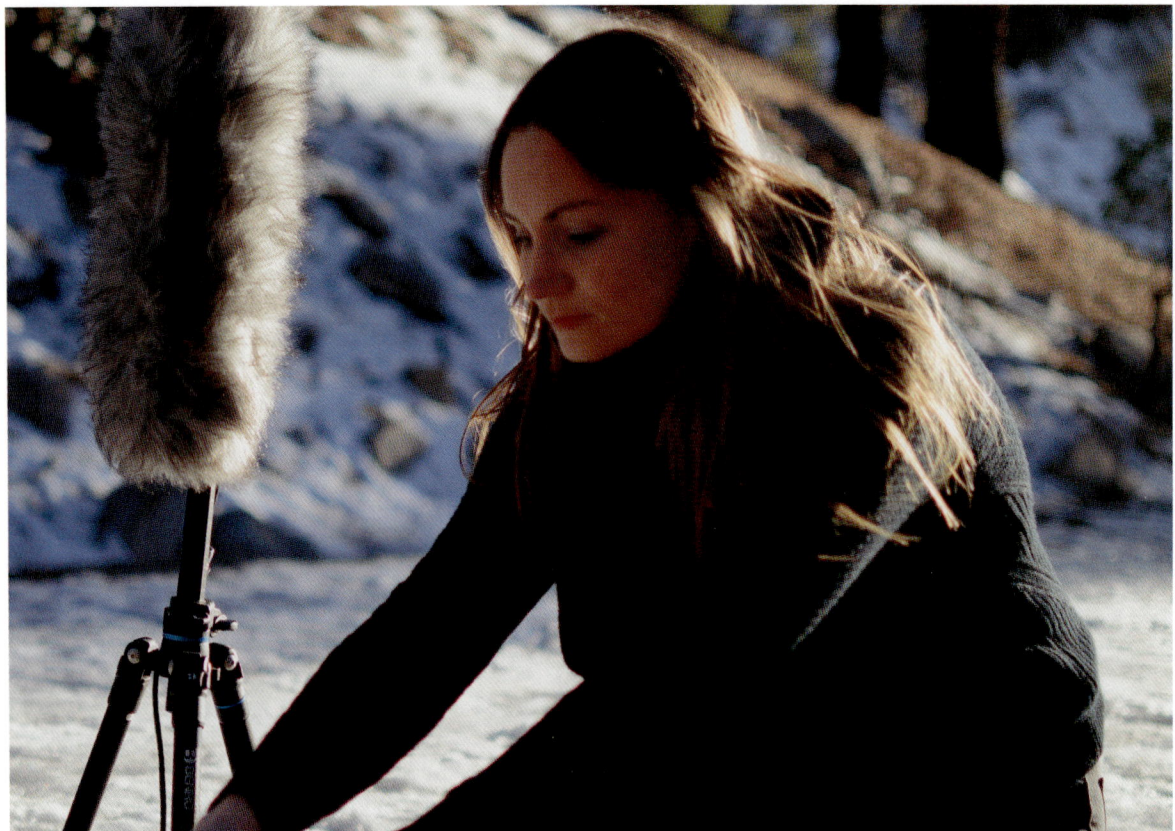

Bethan Kellough recording in the field.
Photograph by Robert Kellough.

RACHEL SHEARER

Based in Aotearoa (the Māori-language name for New Zealand), Rachel Shearer is an artist and composer who affiliates to Rongowhakaata, Te Aitanga a Māhaki *iwi* (extended kinship groups) and Pākehā (New Zealand European). Shearer draws on a divergent set of culturally informed practices that speak to her porous, research-led approach to art making. Her work over the past three decades has explored close listening to the earth and its environment.

Drawing deeply on her connection to Māori cultural traditions, Shearer has created installations, albums, sound works, and videos that recognize and reconcile the Māori concept of "shimmer"—the vibrational force at the core of being that she explores in *Whakapapa of Shimmers* (2024). Shearer is interested in the ways energetic knowledge can be framed through action and inaction, in place and time, and she invites viewers to be carried by—and ultimately transformed by—encounters with her work. In her Te Kaiwhakatere (navigator) role as senior lecturer at the Auckland University of Technology, she prompts staff and students to consider what Indigenized mainstream tertiary education might look like. <

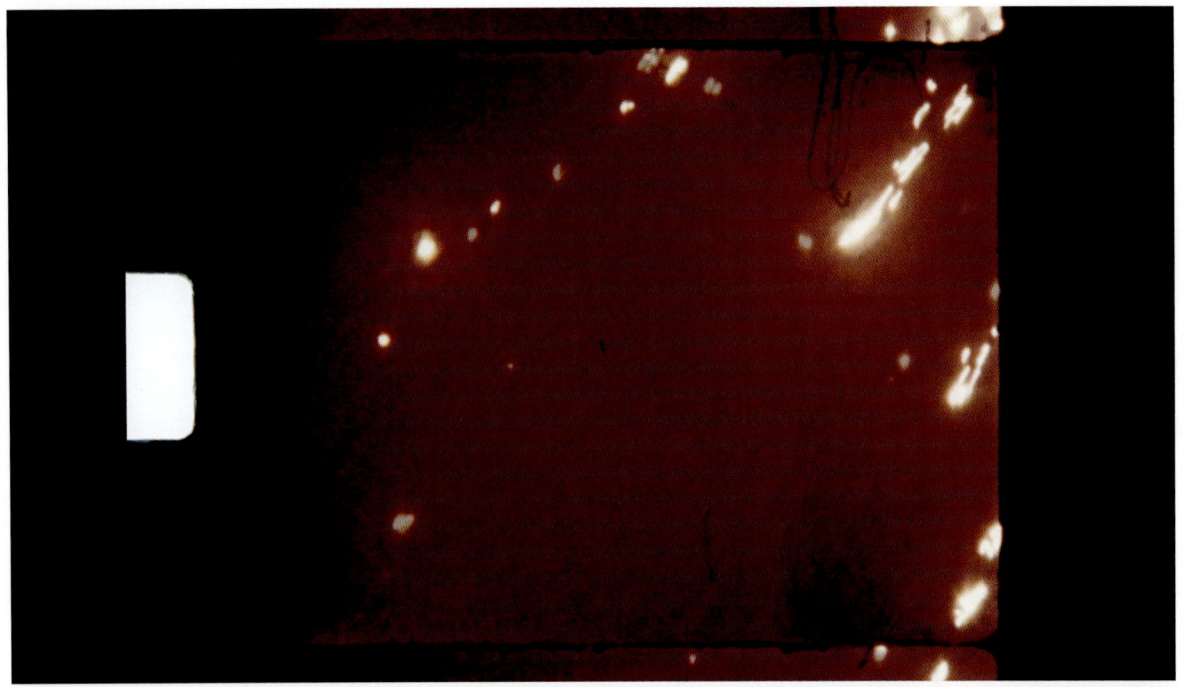

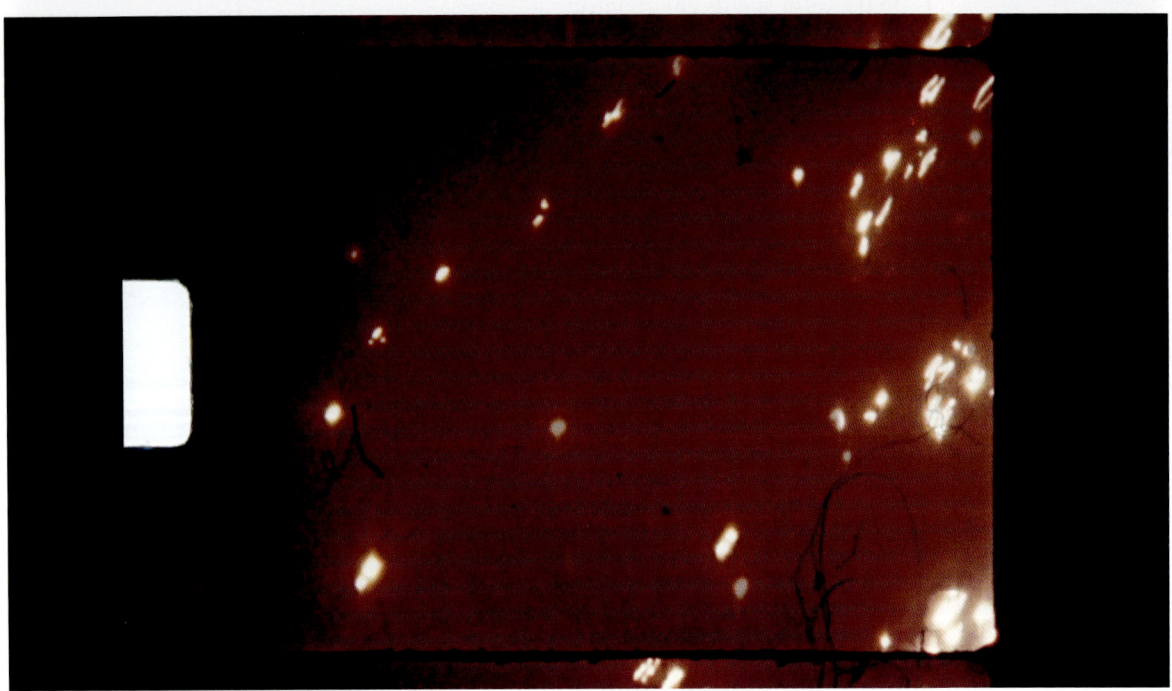

Rachel Shearer, still from *Whakapapa of Shimmers*, 2024. Photographs courtesy the artist.

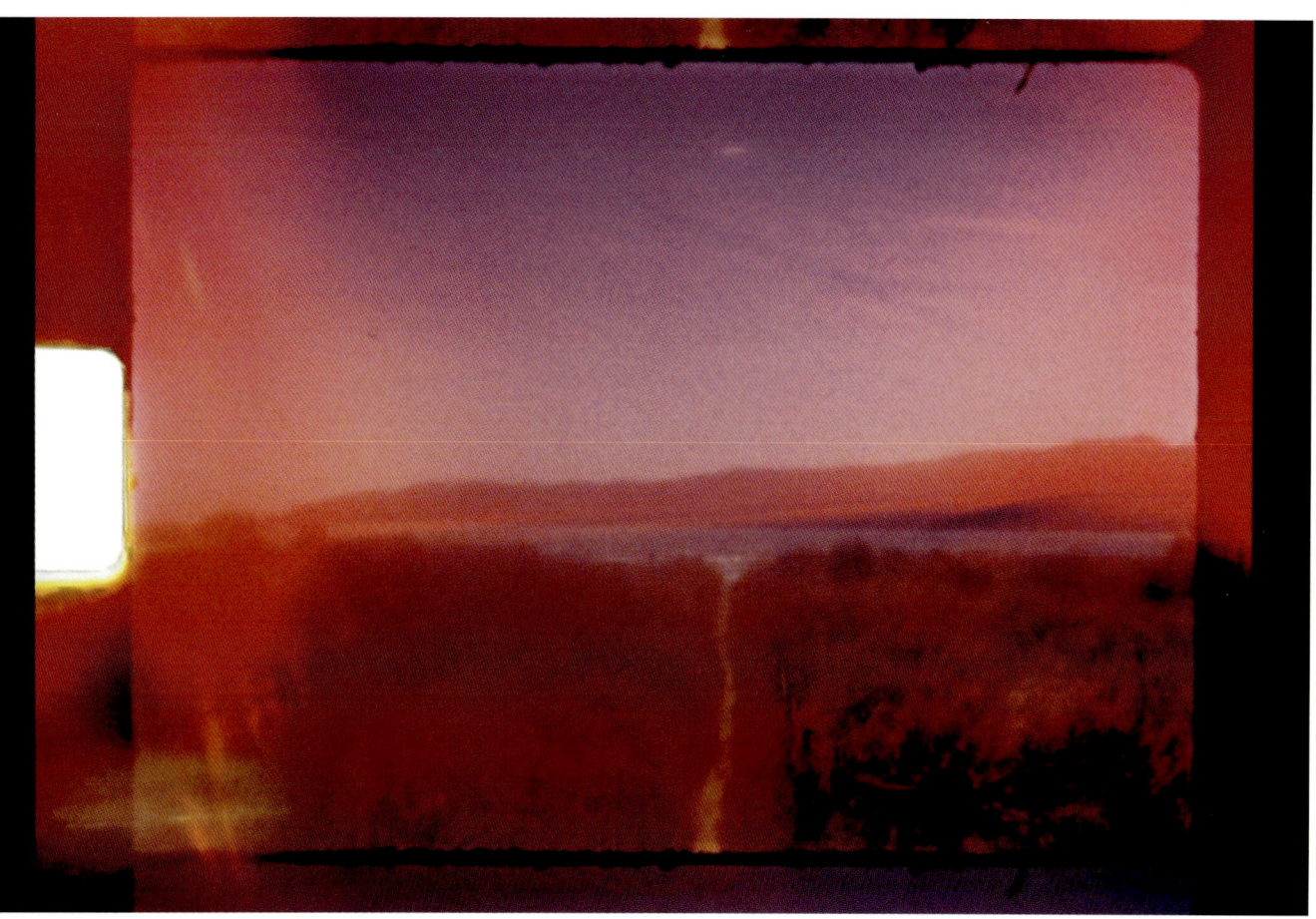

Rachel Shearer, still from *Whakapapa of Shimmers*, 2024. Photographs courtesy the artist.

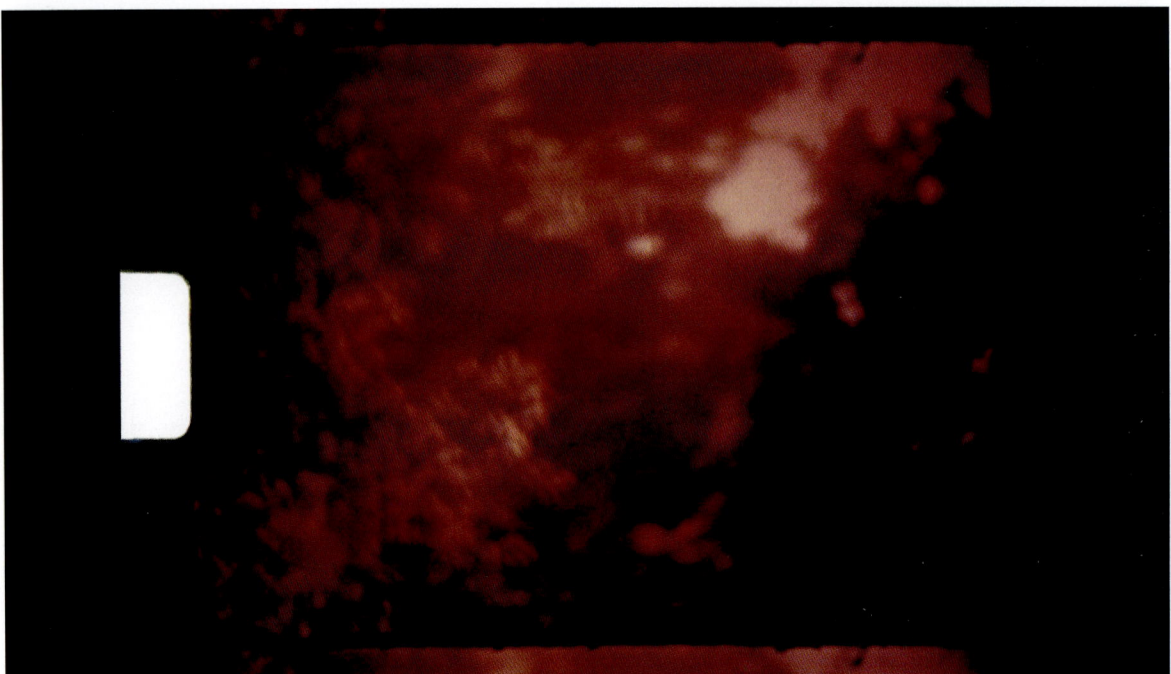

Rachel Shearer, still from *Whakapapa of Shimmers*, 2024. Photographs courtesy the artist.

MINORU SATO

Hailing from Kamakura, Japan, Minoru Sato is an interdisciplinary artist, curator, and musician whose work explores human perception of natural phenomena such as sound and light. His research often draws on principles from the natural sciences, and his installations are evocative of laboratories, where connections between the tangible and the intangible are revealed.

Among Sato's areas of interest is temperature, from the very narrow range within which life can exist to the extremes found in the sun, deep space, and the earth's core. *Energy Fields: Vibrations of the Pacific* presents a new version of Sato's *Thermal Acoustics* (2013–ongoing), which comprises four identical glass tubes heated to different temperatures. As the difference between the tubes' internal temperatures and that of their environment increases, changes in pressure occur, creating audible tones. The frequency of these tones also fluctuates with temperature; the warmer the air, the faster sound waves travel, and vice-versa. By rendering temperature as sound, Sato makes us aware of invisible forces that are shaping both our environment and our perception of our environment. **<**

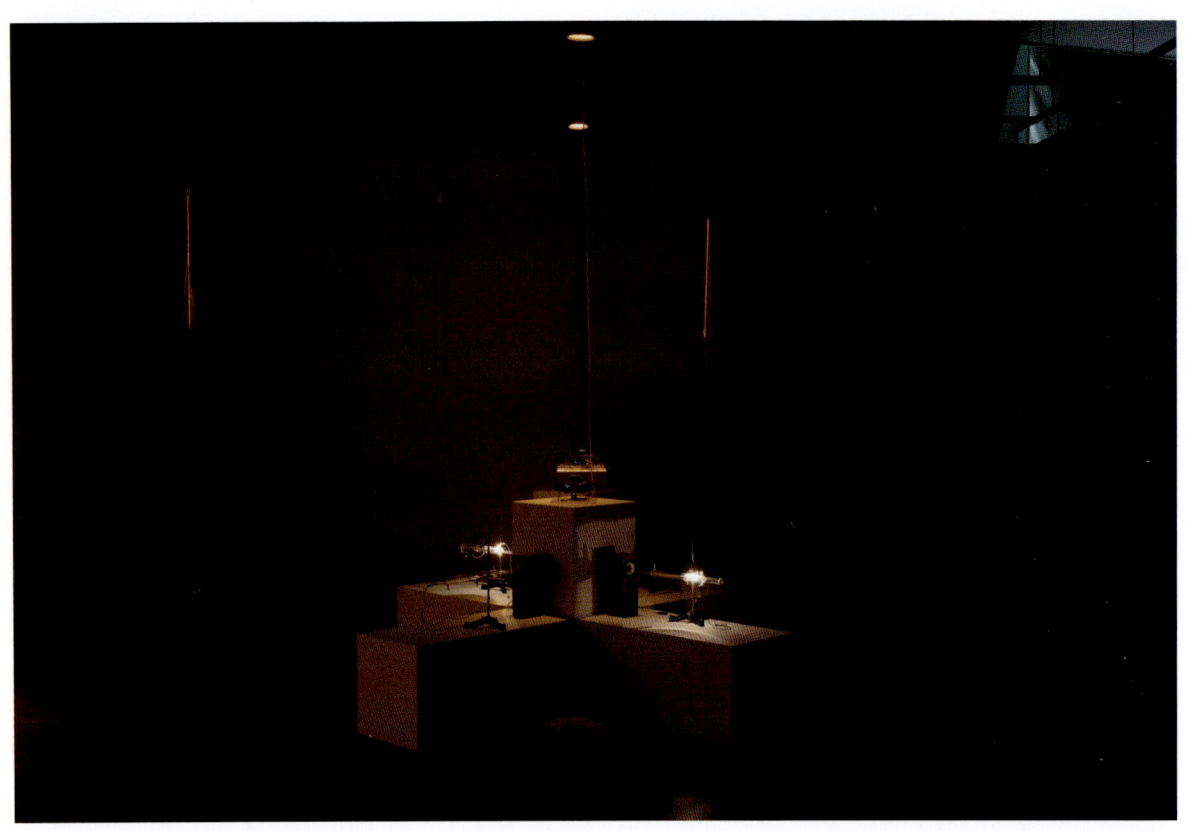

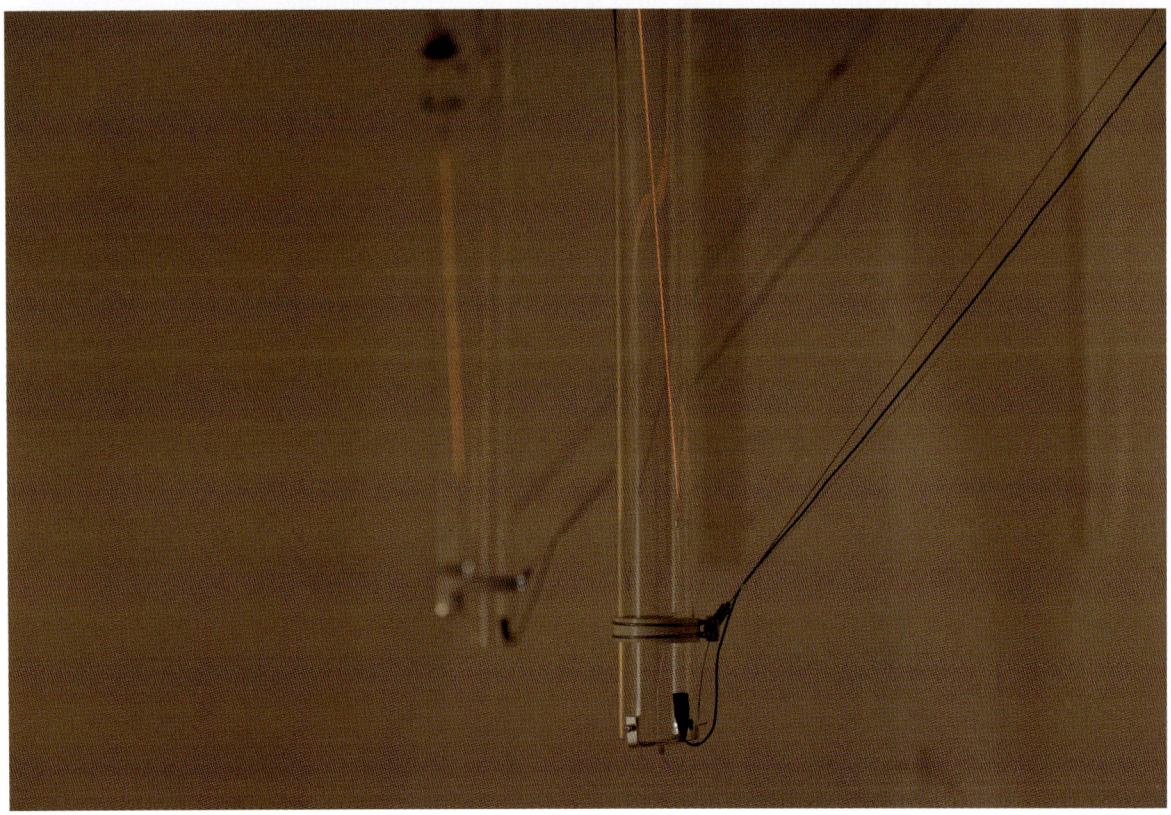

Minoru Sato, *Thermal Acoustics*, 2018. Photographs courtesy the artist.

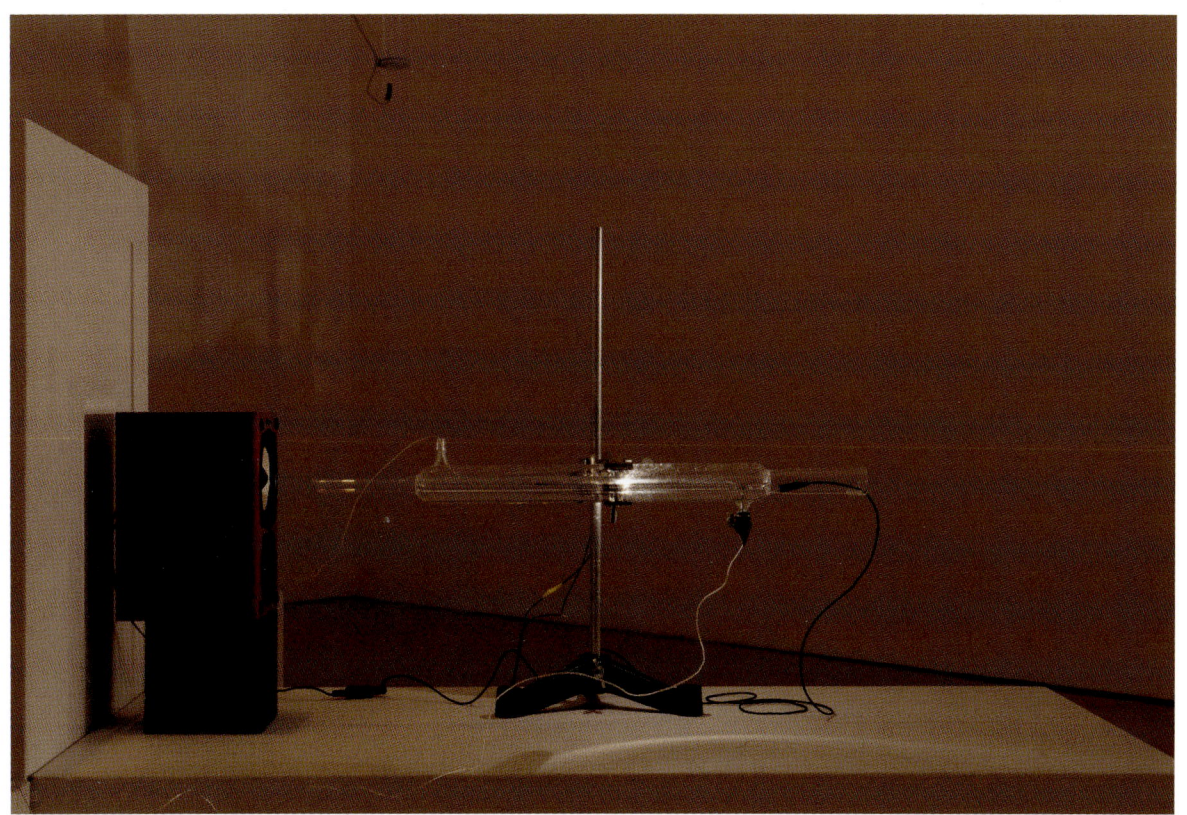

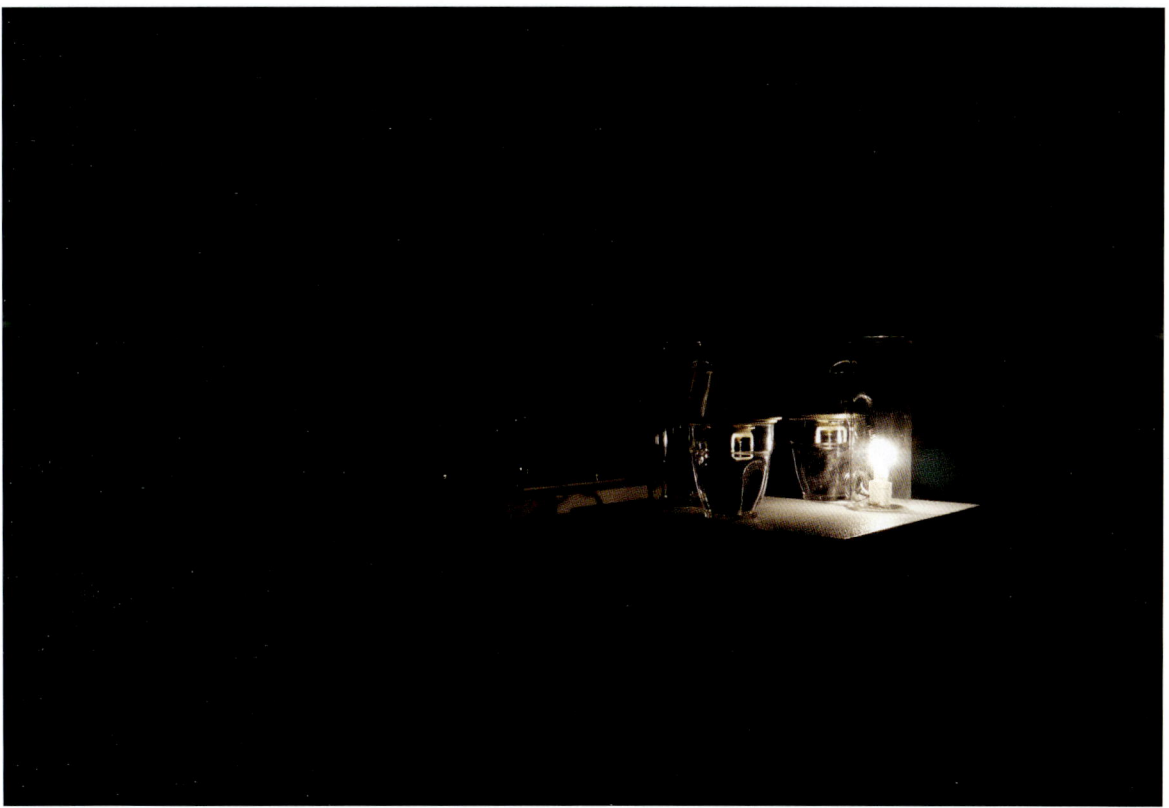

Minoru Sato, *Thermal Acoustics*, 2018. Photographs courtesy the artist.

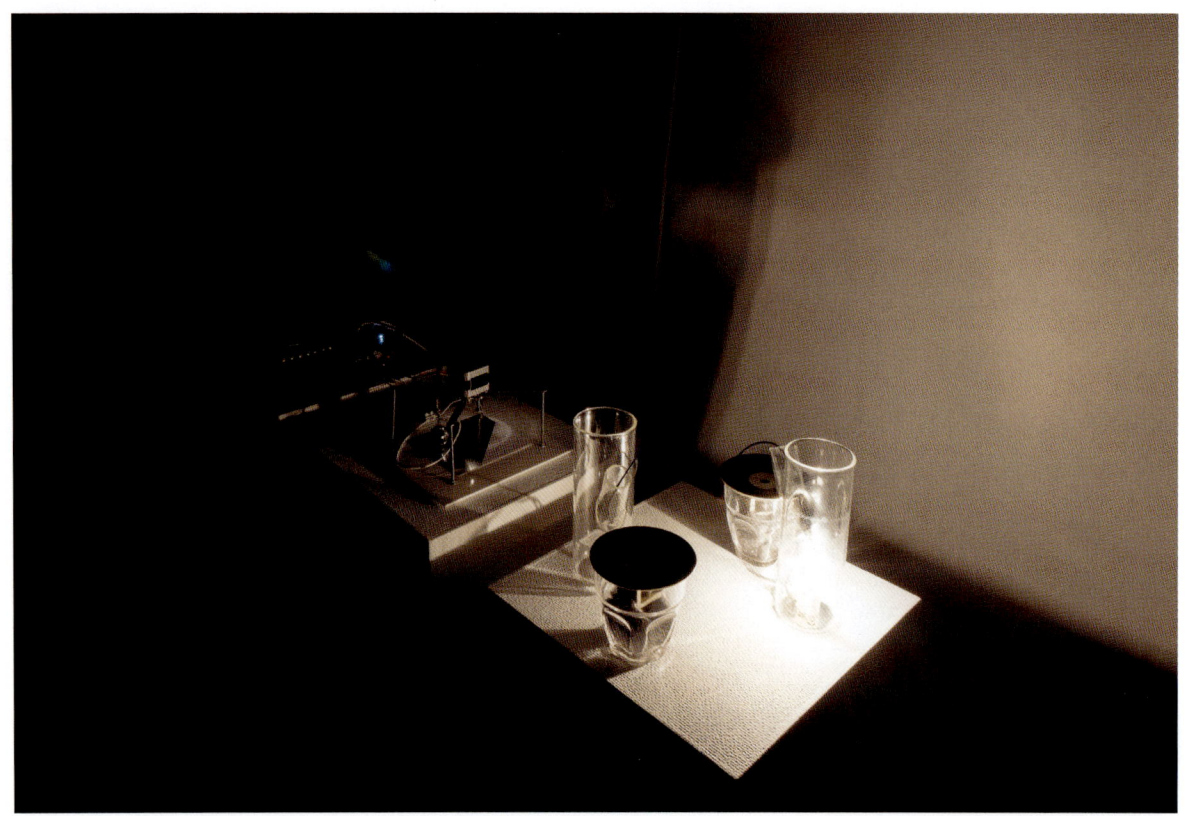

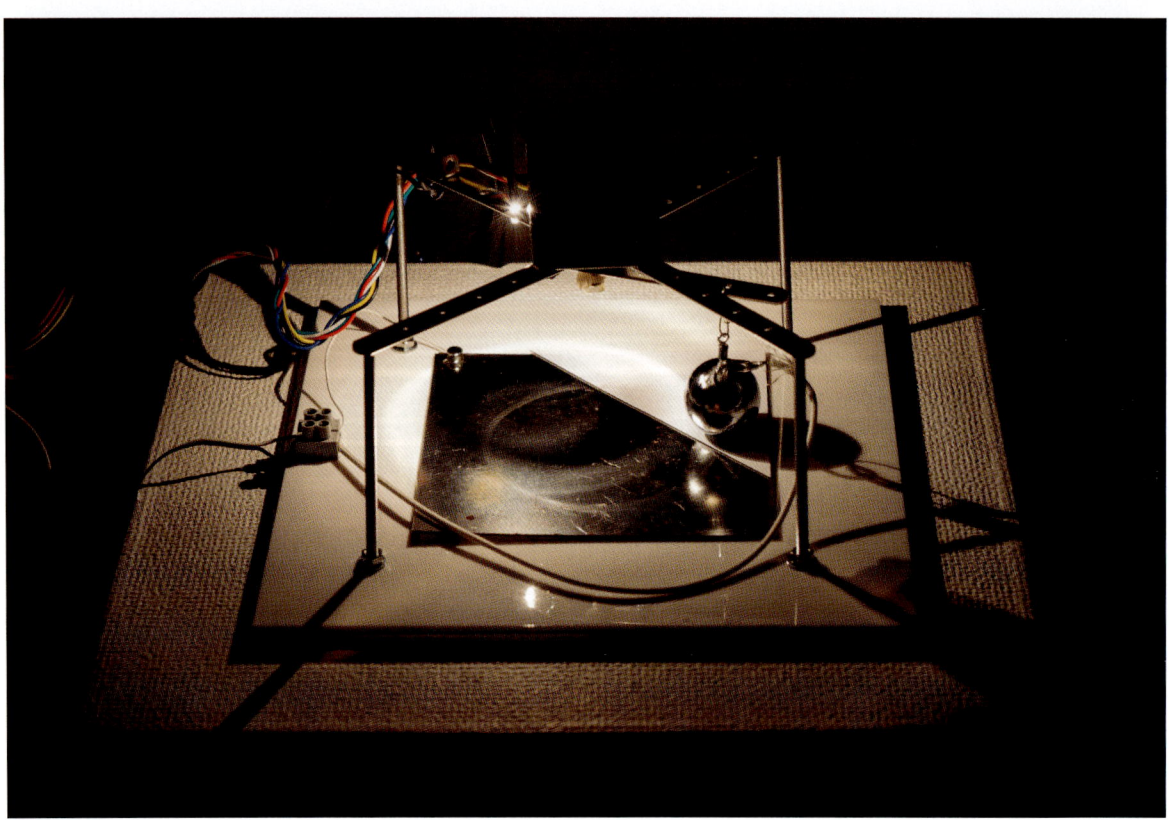

Minoru Sato, *Thermal Acoustics*, 2018. Photographs courtesy the artist.

KYLE SLABB

Kyle Slabb is an artist and cultural leader of the Bundjalung people, the original custodians of the northern coastal zones of New South Wales, Australia. Drawing on different practices, such as painting, music, and installation, Slabb seeks to share ways of knowing that mirror his experiences as well as those of his community. He is an advocate for Indigenous knowledge systems and their application in broader society, and his work explores ways these systems can be shared and deepened.

Slabb's work *Binnangu* (2024) in *Energy Fields: Vibrations of the Pacific* brings visual form to his understanding of country and its continuous imprint on sense and sensemaking. His perception of place is rooted in cultural beliefs that have been passed down to him through elders in his community, and his work reflects an openness to and deep awareness of country as a state of being. <

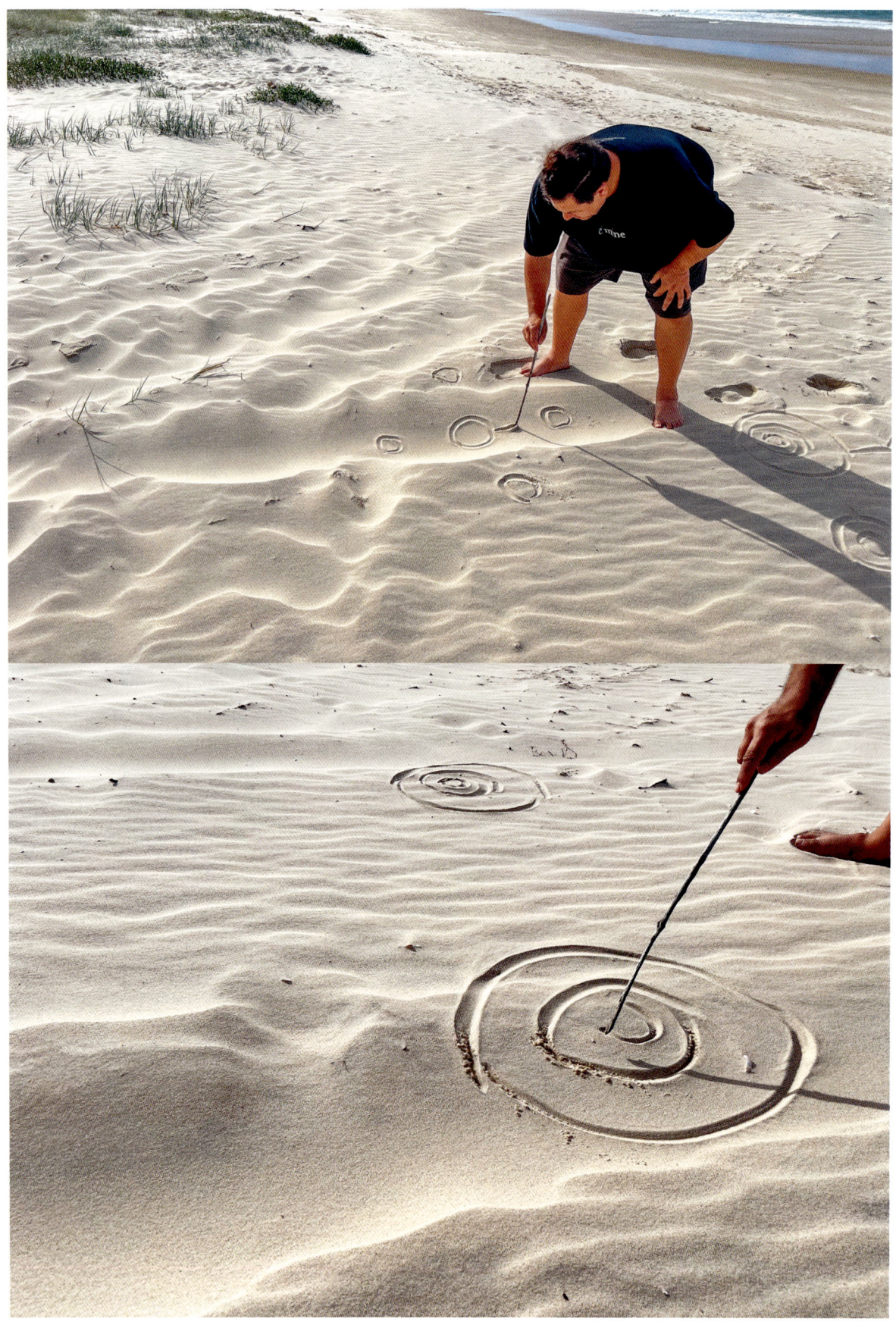

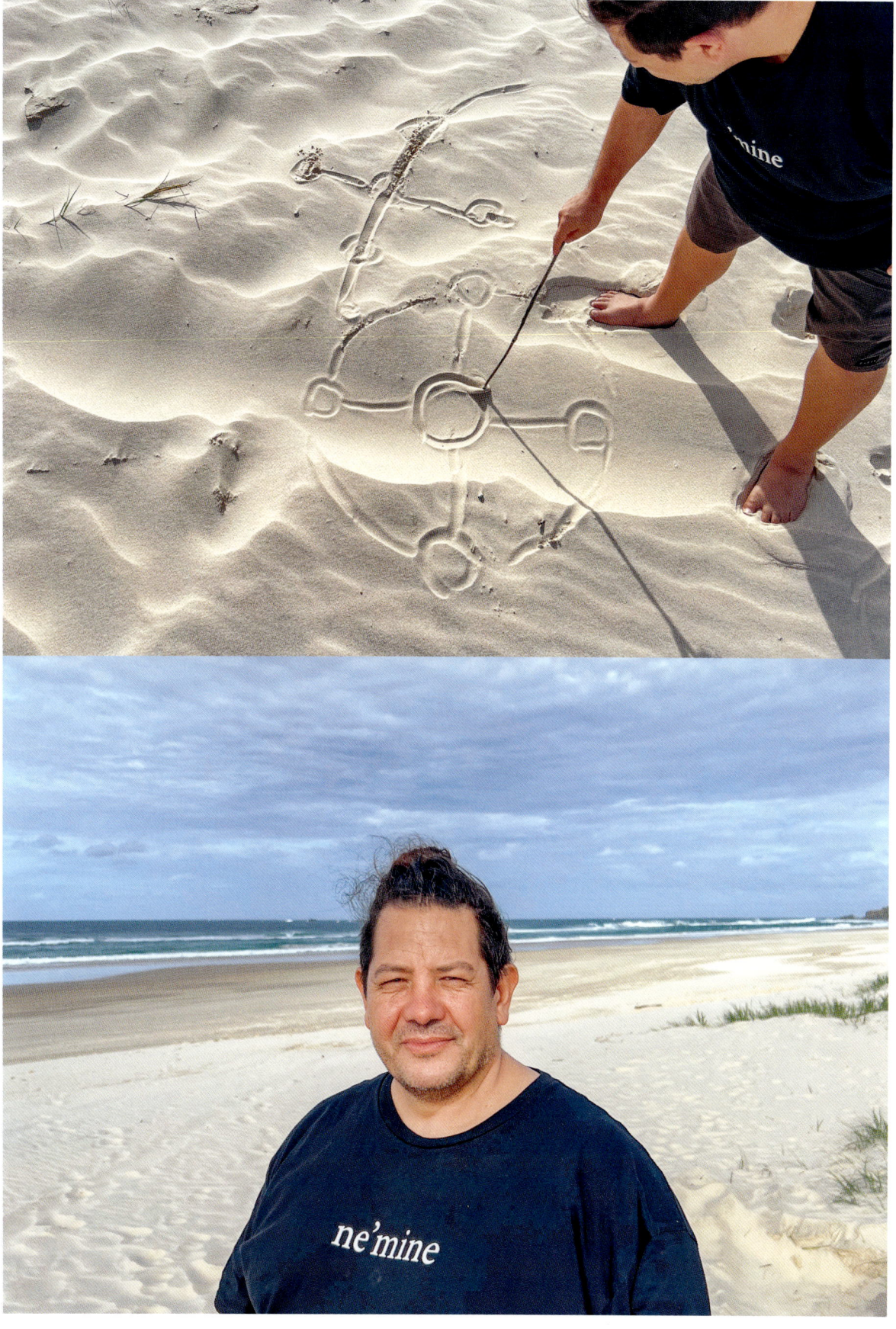

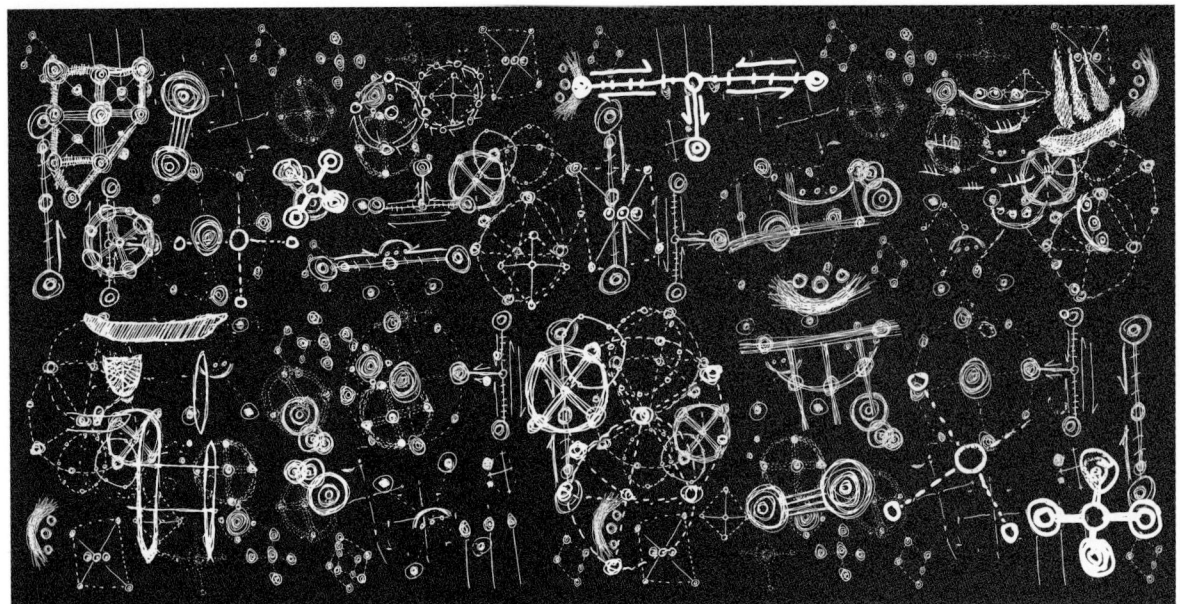

Kyle Slabb, *Binanggu*, 2024. Photograph courtesy the artist.

< previous page spread:

Kyle Slabb, *A starting place of sense, binungal*, 2024. The artist showing the interconnections of land and sky stories at Booningbah (Fingal Head Beach, Australia). Photographs by Lawrence English.

MO
H
ZAREEI

Mo H. Zareei grew up in Ekbatan, a brutalist residential complex developed in the mid-1970s in western Tehran, and his artistic sensibilities were shaped by the stark materialism and geometries of his environment. Zareei's fascination with brutalism is woven through his research and practice, which draws on music, sound, light, and kinetic sculptures, and references both his affinity for minimalism and his educational background in physics.

For the past several years, Zareei has created work in realization of his sound-based brutalism, a concept he developed through practice-based research. *Material Music* (2019 – 20) is a kinetic sound sculpture that exploits the materiality of sound production to highlight the effect of physical material in the transmission of vibrational energy experienced as sound. The sculpture comprises eight pairs of actuators, each programmed to strike various solid blocks of matter, including wood, glass, marble, and copper. Upon activation, the actuators strike the blocks simultaneously, but they slowly fall out of sync; the left and right actuation instances of each unit become offset by minuscule delays driven by variations in the speed of sound as it travels through each material. At times chaotic and other times rhythmic, *Material Music* proposes a potential for a musical language that integrates the phenomenological qualities of matter as an essential consideration for composition. <

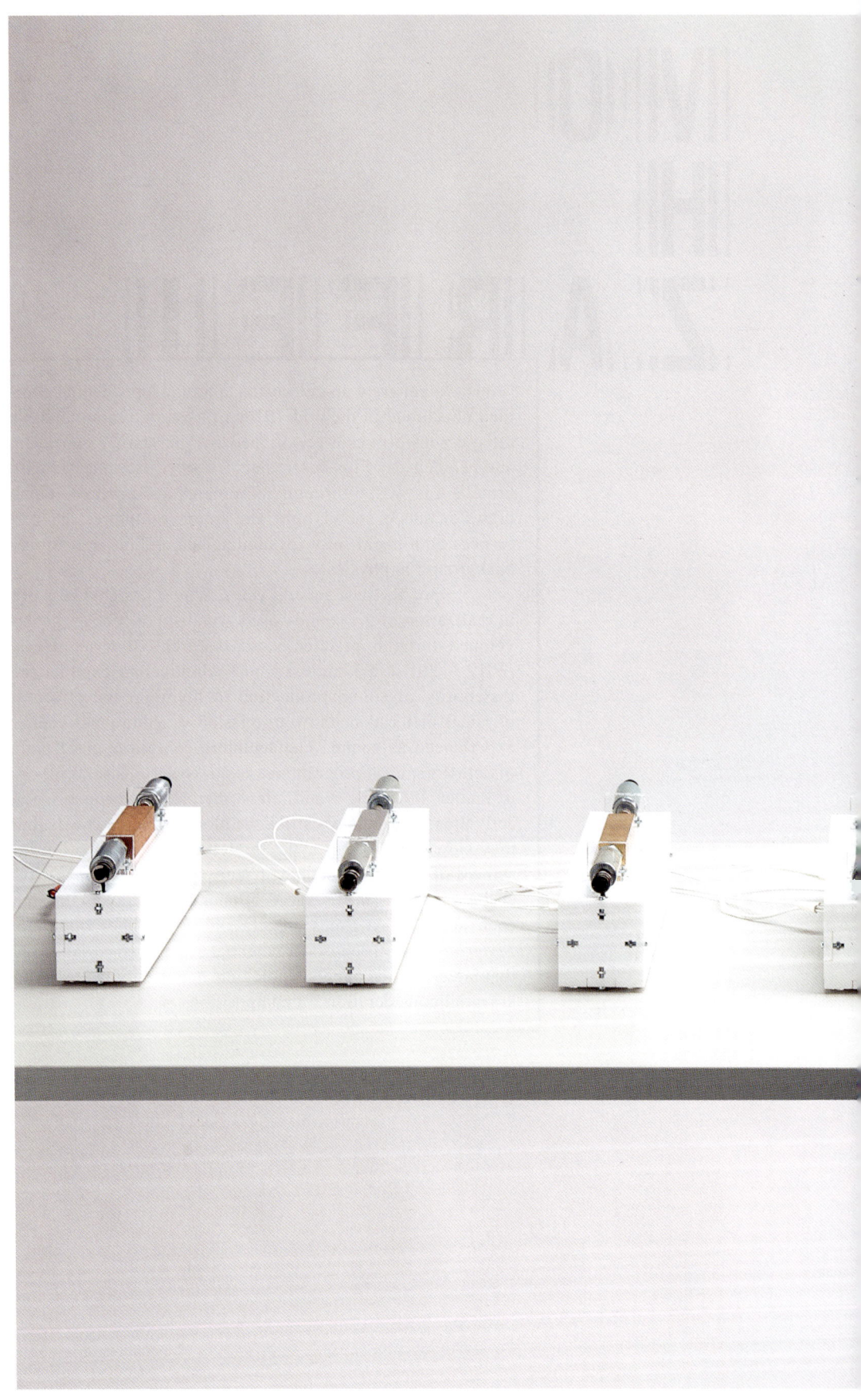

Mo H. Zareei,
Material Music,
2019 – 20.
Photograph by
Gerry Keating,
© Mo H. Zareei.

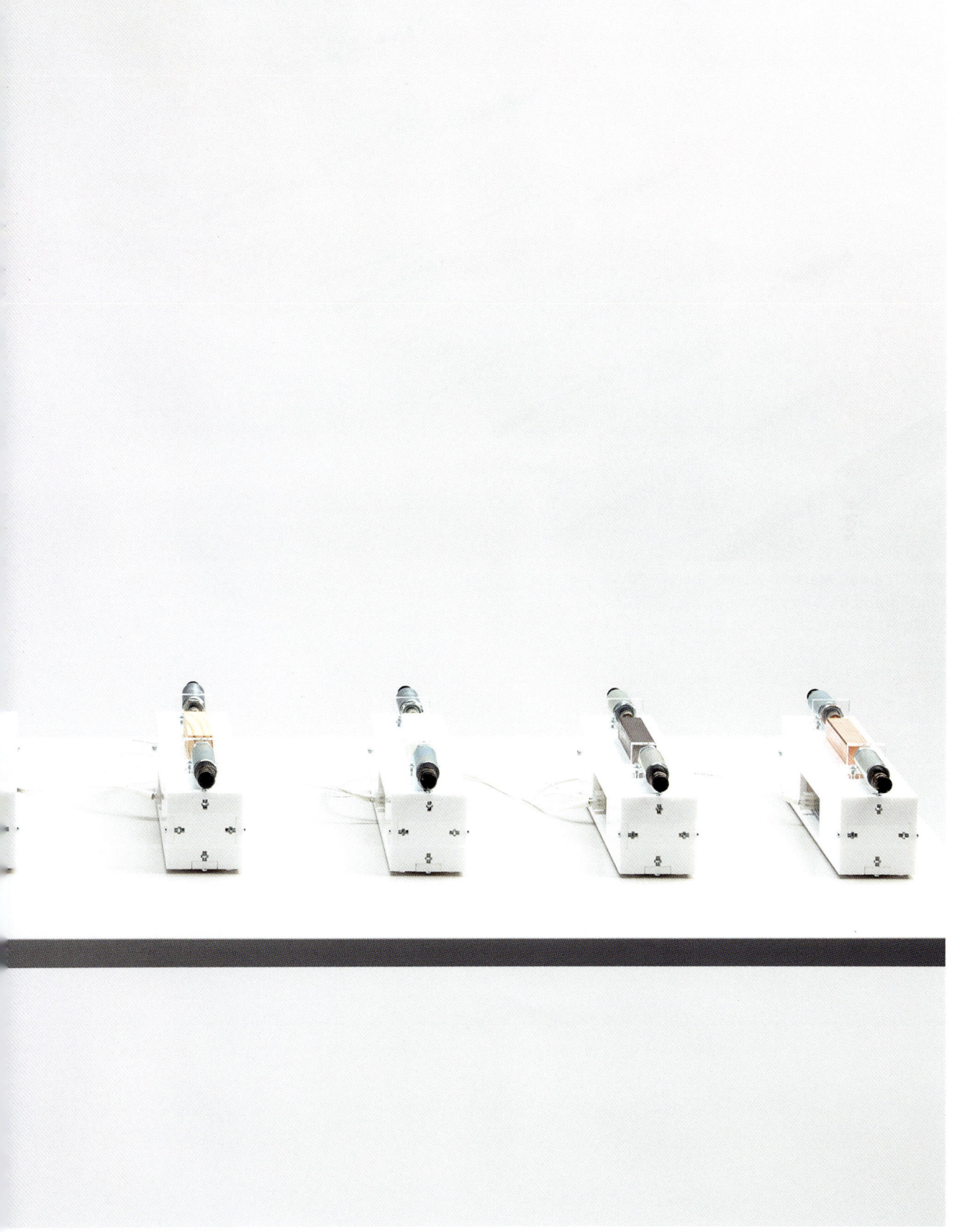

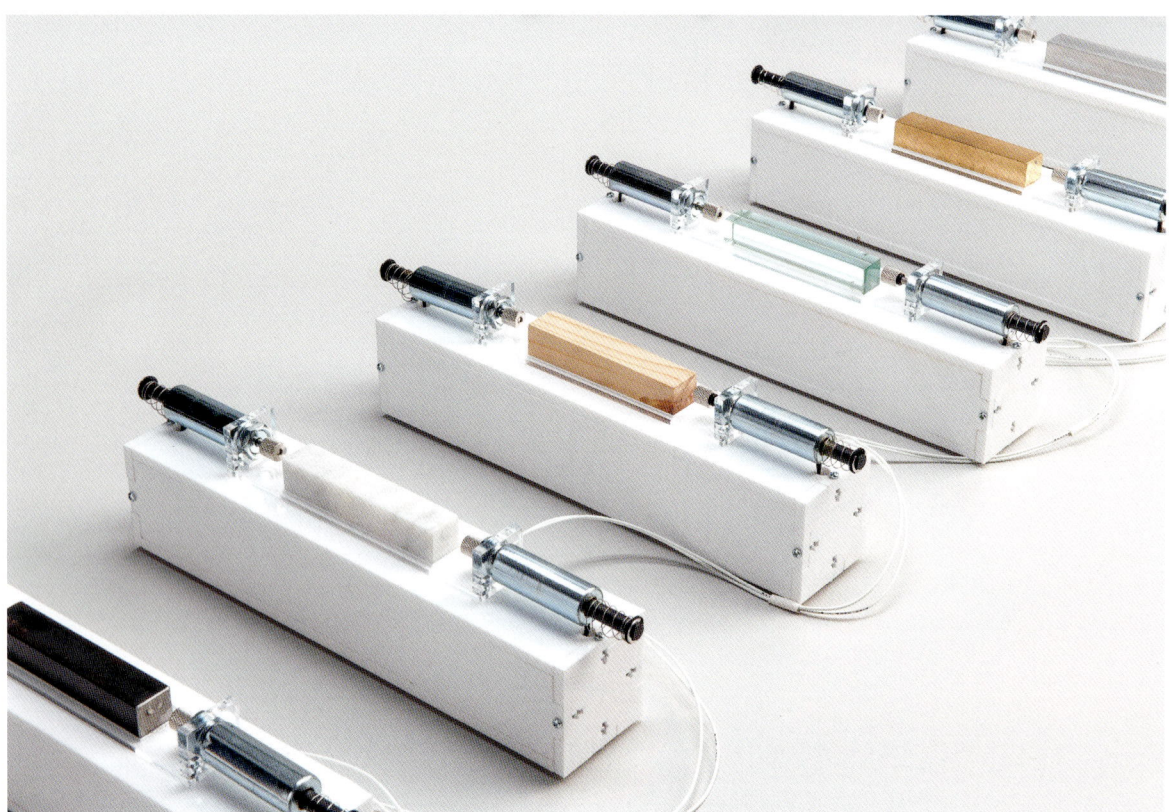

Mo H. Zareei, *Material Music,* 2019 – 20.
Photograph by Gerry Keating,
© Mo H. Zareei.

AKIO SUZUKI

Akio Suzuki's artistic practice has been guided by two words: "throwing" and "following." His methodology, which involves casting energy out into the world and receiving it back, is his way of testing spatiality and materiality and uncovering possibility. In 1982, on the Senrihama Dunes in Hamamatsu, Japan, he made a recording of his self-made instrument, the Analapos, being "played" by the wind. As strong gusts from the Pacific Ocean blew across the instrument, a percussive device comprising two open-ended metal cylinders connected by a spring, it emitted an eerie harmony. The phenomenon prompted Suzuki to move away from playing the Analapos himself, instead using it to translate physical conditions into sounds, striking a dialogue between the dynamism of his environment and the range of the instrument.

In 2023, Suzuki ventured to the Tottori Sand Dunes on the Sea of Japan to again explore the wind's effect on the Analapos. While the core elements of Suzuki's performance—the instrument, the windswept dunes—remained the same, the Tottori Sand Dunes recording captures entirely new sounds, a dialect of wind shaped by energies unique to that specific time and place. Presented in *Energy Fields: Vibrations of the Pacific* as video documentation, the work speaks to the conditional relations between material objects, environmental forces, and perception. <

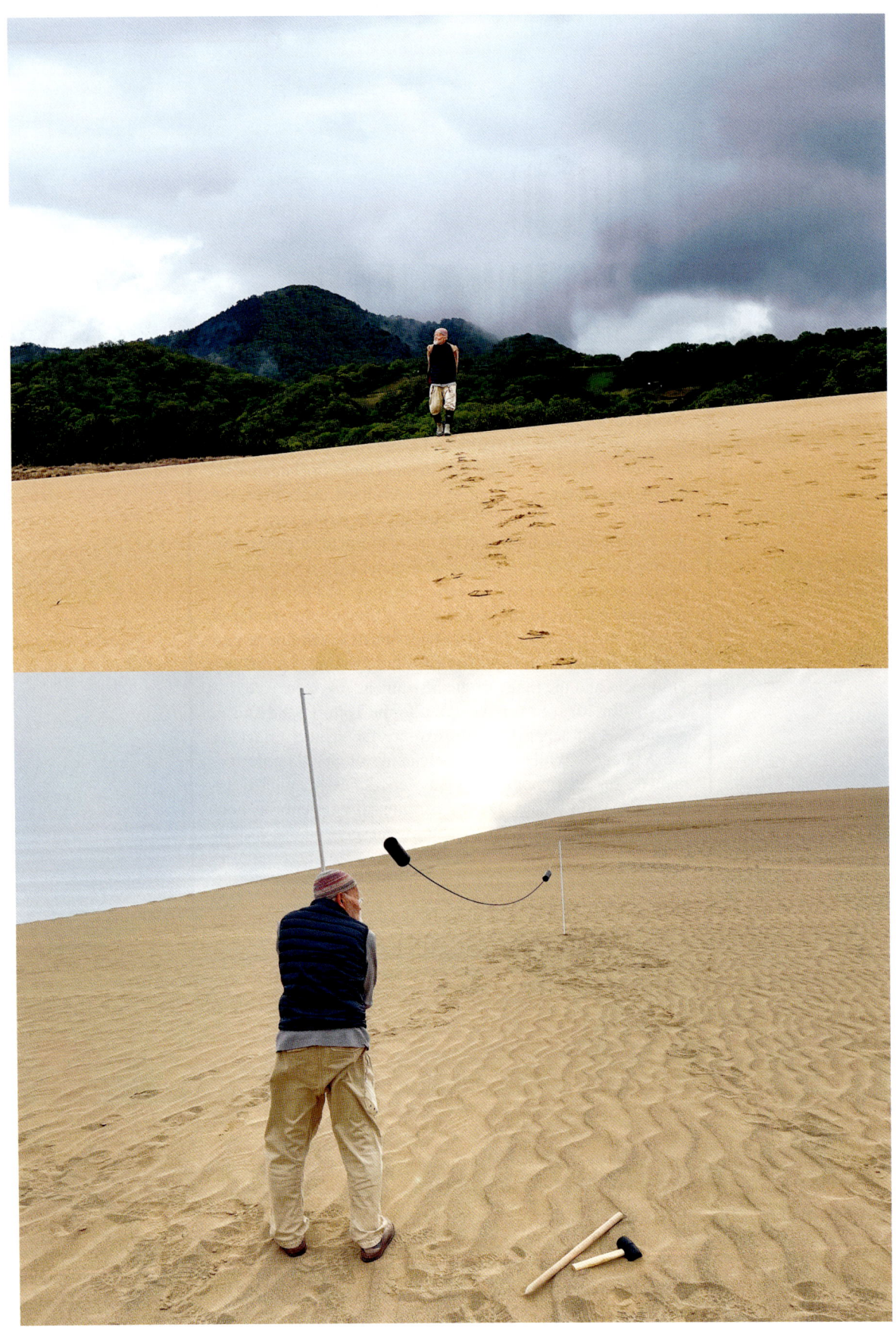

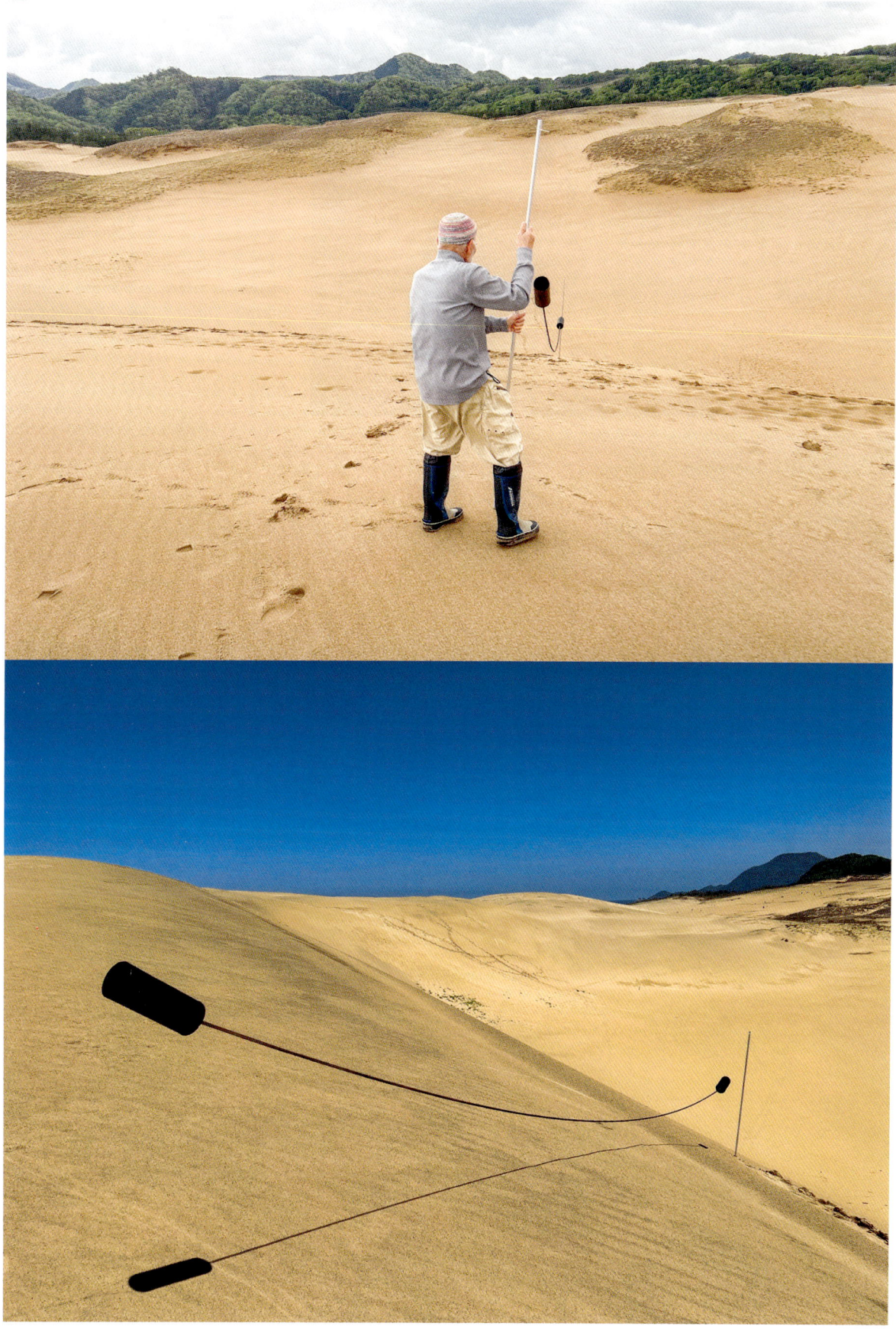

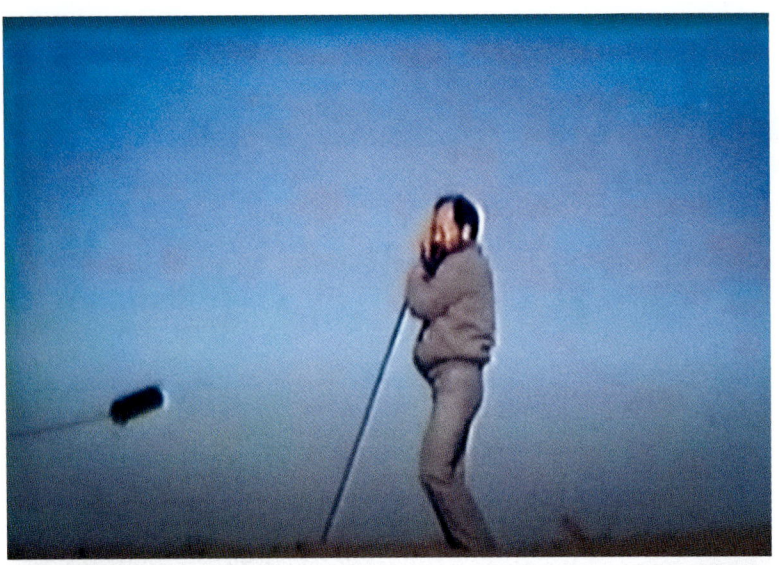

previous spread:

Akio Suzuki with the Analapos at the Tottori Sand Dunes, Tottori, Japan, 2023. Photograph by Lawrence English.

left:

Akio Suzuki with the Analapos at the Senrihama Dunes, Hama-matsu, Japan, 1982. Photograph courtesy the artist.

Akio Suzuki, *Analapos at Tottori*, 2023.

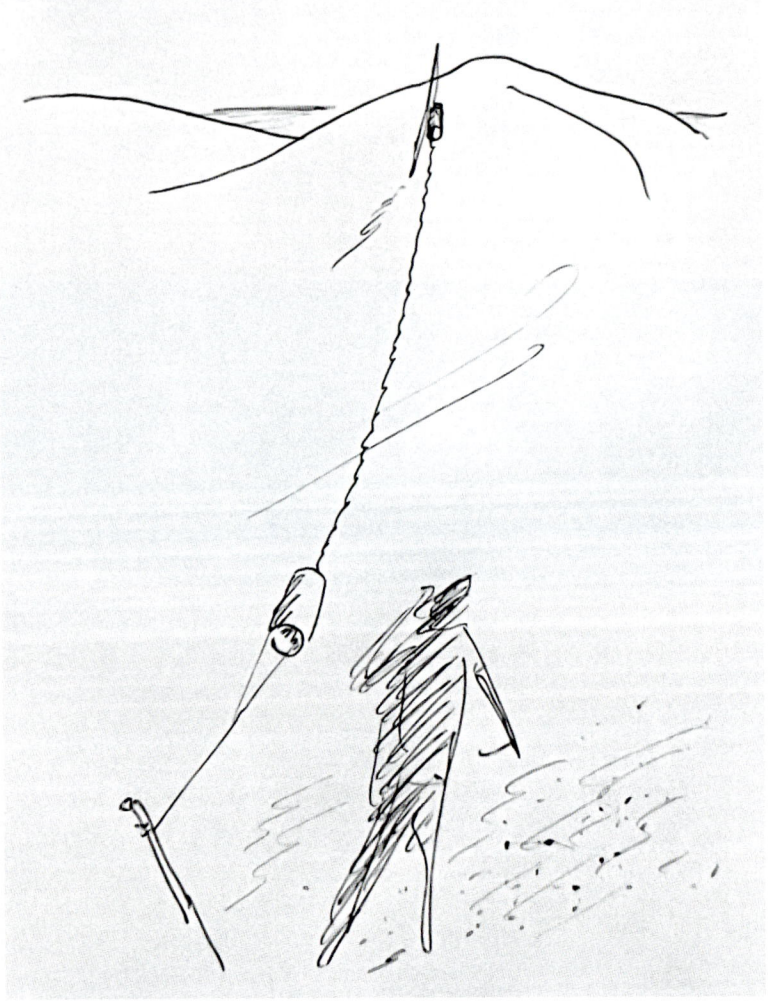

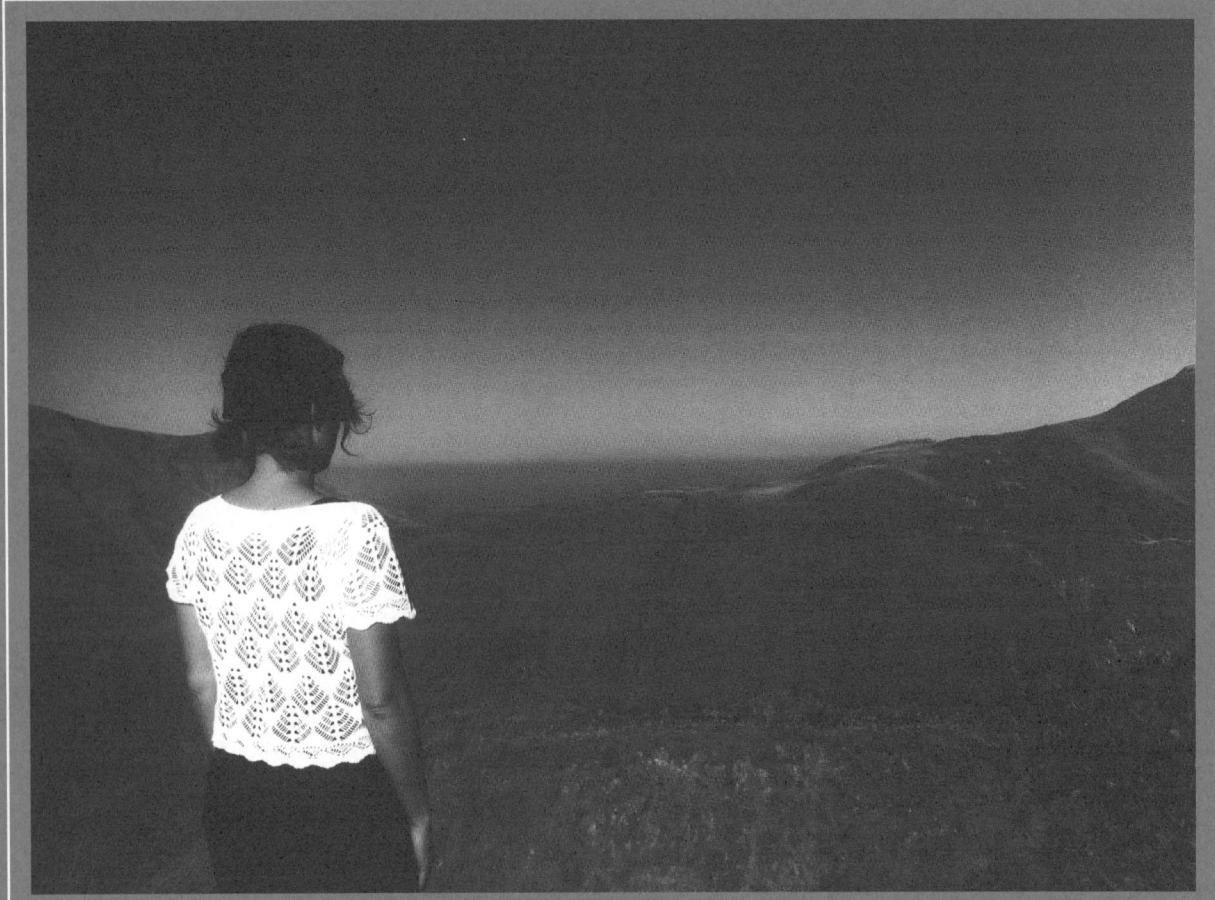

Malena Szlam in the Calchaqui Valley, Argentina, 2015. Photograph by Daïchi Saïto.

LAWRENCE ENGLISH
in conversation with

August 23, 2023

MALENA SZLAM

LAWRENCE ENGLISH: I want to start off by talking about your birthplace, Chile. The landscapes and geology of Chile have informed several of your works, but I sense you have been simultaneously informed by your own personal reflections. You've also done a focused study on the forces that shaped the territories around Santiago, where you grew up. How do you knit your readings of place, the personal, the aesthetic, the geological, and the temporal together? It seems like a complex set of relationships to bring into relief.

MALENA SZLAM: I think these readings are possibilities for me to express views and reveal some comprehension; they create a prism where understanding is unfolding, a perceivable flux that also presents something invisible and inaudible. We can immerse ourselves in a space, our memory becoming a part of it, and simultaneously it can be unfamiliar. I have the curiosity to understand this through the arts and sciences, through cartographies of time, geography, and history. I like that you chose the word "reading," it makes me think of naming an integrative process of perception, translation, and interpretation.

LE: It's a wonderfully porous word, I find. It opens way for the role of the senses.

MS: I agree. And "listening" and "seeing" what the natural world speaks and forms are, for me, thresholds to designing these cartographies using imagination and knowledge. I think of artists and scientists as collectors of imprints and signals, articulating mesmerizing approximations of phenomena and creating artifacts to present and respond in some way to so much that is fascinating and inexplicable. This generates a sense of discovery. This sensorial and cognitive experience develops and adapts in connection with how our technologies evolve and the interference of humans in ecosystems. When we share and learn from each other, we have this amazing opportunity to change the ways we experience and how we see ourselves. My understanding can evolve when I am open to a process of discovery. The body learns to create through those trajectories, aware or not of them, in a process of adaptation to all sorts of environments, artificial and natural. I search for new modes of interaction when making new work, particularly when filming, allowing and falling into a flow.

LE: The body is also situated in a broader context, too. Personal, social, and political histories all have an influence on that idea of trajectory.

MS: Yes. In that universe of reflections, I realized that coming from a nomadic family from diverse geographic and cultural origins and in exile has brought an urge for land. The perspective of seeing this as being unrooted and displaced became a need to understand the connection with my Aboriginal ancestry from the Calchaquí Valleys, to think of their extinct language and how the culture in the valley continued. I became a nomad myself, and part of me settled in Montreal, a landscape so different from the one I grew up in. Moving back and forth from the southern to the northern skies, I searched for roots in the Andes Mountain range, growing and gravitating within its geology and worldview. The space of the mountain, I believe, has defined us as Andean people, recovering a connective identity beyond the geopolitics of borders. Seeing each other not as Chilean, Argentinean, Peruvian, or Bolivian, but as Andean peoples.

This geography of folding mountains—the image of a *puchu* in Quechua or *pontro* in Mapudungun or *poncho* in Spanish—draws and sculpts the valleys from the altitude where glaciers live to rivers forming and flowing towards the Pacific and Atlantic Oceans. The slope of the mountain range, as sinuous movement, is the arc connecting the east and west coasts below the Southern Cross. The Pacific Ocean transcends borders, with flora and fauna that are familiar from the southern to the northern hemisphere. The strata and minerals are the architecture of the earth's crust, its interstitial flesh, crystals and volcanic rocks assembling temporalities of massive-scale events. Rocks contract and expand, alive even when we perceive them as quiet, inanimate, or still, because their crystals are growing and transforming in time. I imagine how small and large disturbances in the geologic structures are shaping the earth.

LE: I feel that's reflected in how your work is both conceptualized and presented.

MS: When I think about how a project begins, it's like thinking about how an experience begins. There is not a single moment, but rather trails—a constellation of possibilities—and the conviction and perseverance to allow processes to unfold and open to new cycles. Some of the stories behind the works are what shaped the decisions and impulses to search for and live experiences. At some point, the life of the project takes form on its own, and as an observer, first, then as a maker, I keep strengthening that process.

LE: I recall reading comments about *Altiplano* [2018] that described a kind of "vibrating landscape" quality to the film. For me, though, I got a sense you were reaching into the landscapes, reaching into time, and trying to reflect on the nature of vibration not just in a geophonic sense, but also in terms of the environment. The work is a sort of compositional score that reflects many hundreds of millions of years of pressure being exerted on those locations.

MS: I find joy in exploring and observing the natural environment, how it manifests space and time converging multiform. Through the experience of filming, I am mesmerized by light, colors, and textures. I am holding onto a bit of that atmosphere in the film's surface by weaving in fractions of time and forms that become rhythms evolving in my mind. It is a beautiful feeling of freedom that can also disappear or be taken away. I have been thinking about how special the moment of filming is, how fragile it can be as well, and how intimate.

LE: As an artist and filmmaker, I am interested to understand how you explore and make tangible these ideas around vibrations, time, and the landscapes that are the material manifestation of energetic reactions. What role do scientific methods play in your work? How is it you can translate some of these quantifiable conditions into more qualitative expressions?

MS: My materials are color reversal film and a 16mm Bolex camera. I see my filming in part as trying to translate language from environments, approximating those voices, intertwining them with interpretation so they remain in partial sight. There is also a desire to treasure these moments. The visual vibration, for me, becomes that existing ephemeral time and space that fade as it is held. In each image, every single frame translates trails and echoes of color temperatures like air and atmosphere. It is a form of memory, permeating and growing cellularly in the terrestrial and celestial bodies as well as on the film surface that is light sensitive. Attending to the sensory geography of the earth and sky, the film layers latent images from diverse temporal and spatial qualities to imagine and get closer to the physical processes of the earth.

With *Altiplano*, I integrated sound for the first time; until then, my films were all "silent." I was in the process of viewing the 16mm footage I had filmed in the Altiplano region and thinking about the sound for the film when I met Susannah Buchan, an oceanographer, in Chañaral de Aceituno in Chile. She was doing field research on the ecology and acoustics of blue and humpback whales in the Humboldt Archipelago. This encounter opened a universe I never imagined before, as I became aware of the spectrum of sound we cannot hear. Months later, I met geoscientist and filmmaker Clive Oppenheimer, and my understanding about signals and language felt deeper. For *Altiplano*, I organically became a collector of sounds, and with these materials I developed a soundscape combining infrasonic recordings of whale vocalizations registered by Susannah and tectonic activity at Mount Erebus recorded by Clive, intermixed with field recordings of lava flows, wind, fire, heartbeat, and spouting geysers. The final sound design and mix was performed in collaboration with James Benjamin and Mamo Koba in Montreal.

LE: How did these sonic materials connect for you within the work?

MS: They connected through conversations about infrasound and learning that the population of blue whales in the southern Pacific Ocean have developed their own unique dialect. By differentiating between whale songs, scientists can understand their migratory journeys and develop conservation planning. Whales are the heart and brain of the ocean. There is less light in the ocean, and water is denser than air, so sound is more essential for communication than vision.

I am curious about inaudible frequencies; they surround and influence us as well as tell us about the early formations in the Altiplano region. Over millions of years of evolution, rocks and mountains emerged and evolved, affected by seismic movements and wind patterns, and species migrated and were born. Another fascination is altitude and its effect on our bodies, gravitational forces and gravitational time dilation. According to the theory of relativity, time moves faster at altitude. <

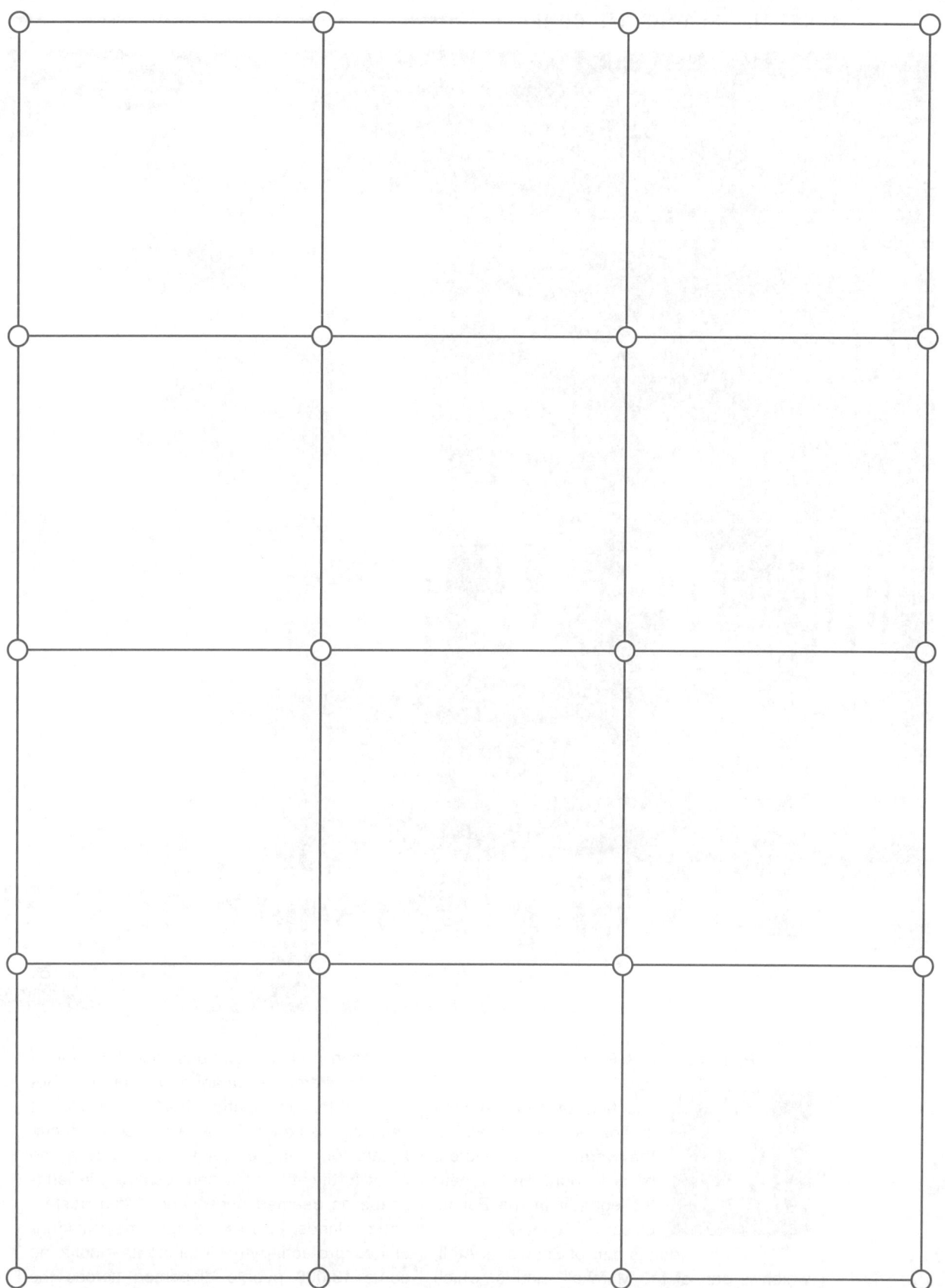

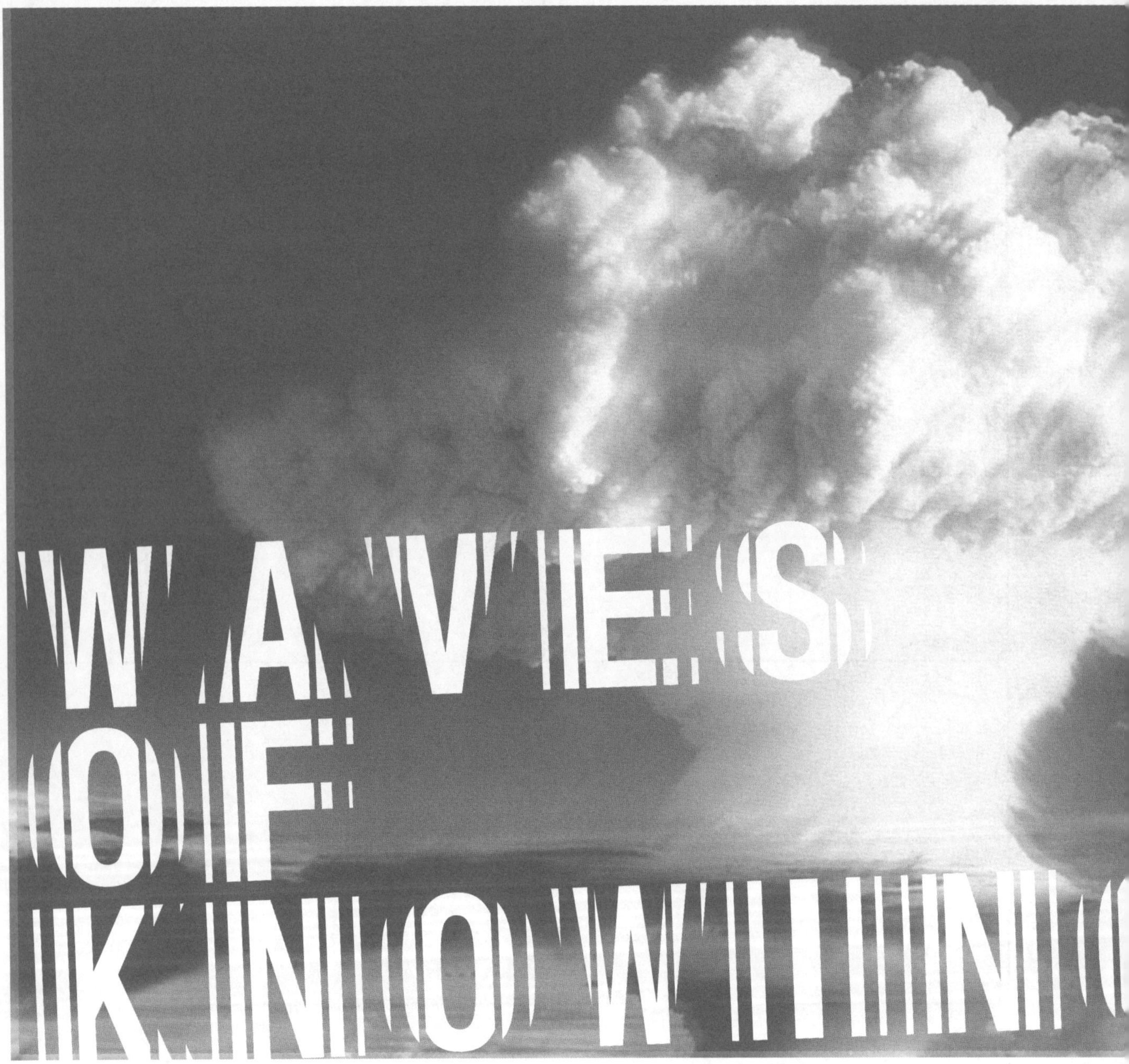

WAVES OF KNOWI

by W. PATRICK McCRAY

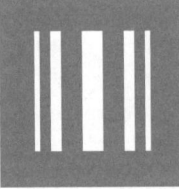t is May of 1951. A man is floating face down in the warm shallow waters of a lagoon. He is near the equator in the Pacific Ocean.

The lagoon is part of Enewetak Atoll, which is part of the Marshall Islands, which is part of Micronesia. Since 1947, Micronesia has been a United Nations trust territory—a "strategic area"—managed by the U.S. Because of this designation, the U.S. has almost limitless discretion for what it can do there. It can, for example, build military bases and erect fortifications.[1] It can reshape the landscape as deemed necessary. It can destroy entire islands. Part of the agreement also includes protecting the inhabitants—including some 13,000 people dispersed across the twenty-nine islands and atolls, some seventy

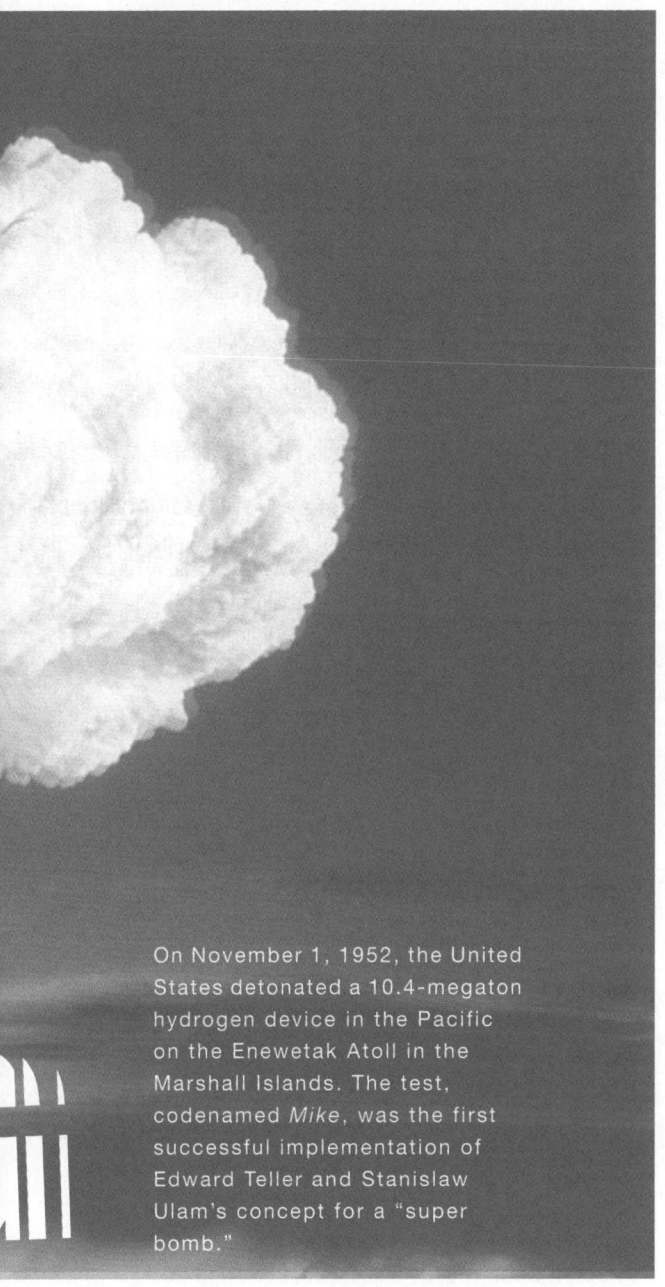

On November 1, 1952, the United States detonated a 10.4-megaton hydrogen device in the Pacific on the Enewetak Atoll in the Marshall Islands. The test, codenamed *Mike*, was the first successful implementation of Edward Teller and Stanislaw Ulam's concept for a "super bomb."

finds the contrast between the calm and the turbulence remarkable. He puts on a snorkel mask and watches the fish below. They are, he will write his wife, "like birds in a tree," but even "more colorful, more numerous, and, to human ears, completely silent."[3]

In a few days, he will help obliterate part of Enewetak.

Three hundred miles east, on the atoll of Rongelap, there is a boy. He is only three years old. In May 1951, he is not thinking about waves. But he soon will be. Like the physicist, he will look at the waves and reflect on them. In time, he will learn how to use the waves to know things. They will help him learn where he is on the open ocean. He will become a navigator. But before that will happen—like the physicist used to be, before he became a physicist—the boy will become a refugee. He will be displaced from his home because the physicist helped destroy it. Years later, when he returns, much of what he once remembered will be gone, lost, erased.

If we travel east across 5,000 more miles of ocean, in May 1951 we encounter another man, working on the coast of Southern California. Like the physicist, he left his home in Europe years ago to escape political chaos. He is an oceanographer. Like the boy, he spends much of his life on ships, studying the ocean, learning from its waves, and trying to understand them. In 1952, the oceanographer will visit Enewetak and help test an idea the physicist has proposed. This test and many more like it will gradually erase more and more of the boy's homeland. Later, when such ex-

square miles of land in all, that make up the Marshall Islands—"against the loss of their lands and resources."[2] The U.S. is not doing this last job well.

The man in the lagoon is not thinking of legal agreements or moral responsibilities. He is a physicist and, right now, he is thinking about waves. The lagoon is peaceful, a smooth lake of ocean water protected by a chain of small islands. Outside, on the coral reefs, larger waves break and boom. He

1. "Trusteeship Agreement for the Former Japanese Mandated Islands, Agreement Approved by the Security Council of the United Nations, 2 April 1947," *Department of State Publication* 2992 (Washington, D.C., 1948).

2. Ibid., Article 6.

3. Edward Teller and Judith Schoolery, *Memoirs: A Twentieth-Century Journey in Science and Politics* (New York: Basic Books, 2001), 322.

periments are no longer done, the oceanographer will create a new way to learn from the ocean's waves. He will use them to measure a new risk to the home of the boy, who has since grown up and become a navigator of ships. This new hazard might one day make the navigator's home island disappear forever, lost beneath the waves.

***

These three people—Hungarian-born physicist Edward Teller (1908–2003), Marshallese ship captain and navigator Korent Joel (1948–2017), and Austrian-born oceanographer Walter Munk (1917–2019)—never encountered one another in the Marshall Islands. Two of them were famous scientists, feted by presidents and decorated with many prizes and awards. The other was unknown except in his own local community of watermen and navigators. But their lives intersected in oblique ways, creating ripples and eddies around events occurring at different times and places. This essay explores a series of episodic resonances and reverberations that emerged from these intersections. It connects Cold War–era research in the Pacific with the nuclear and environmental displacements experienced by the Marshallese. It places traditional Western methods of scientific understanding in conversation with indigenous and local knowledge.

My three main historical characters are linked together by a fourth actor: waves. Whether seismic, sound, or oceanic disturbances, their existence and essence was never in doubt. This is not an ontological inqui-

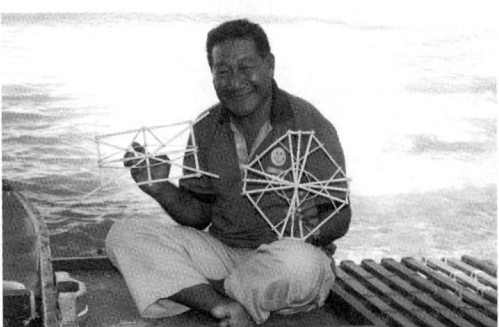

Captain Korent Joel with stick charts, Majuro Atoll, Marshall Islands, 2006. Photograph by Joe Genz.

ry but rather an epistemological one. I wish to ask what messages the ocean's waves transmitted at various times to Teller, Joel, and Munk. I explore how these oscillations communicated and conveyed local and global information, sometimes across vast distances. As waves of knowing, these different vibrational episodes compel us to contemplate diverse ways of understanding the world. Waves of knowing were also, in other words, ways of knowing. Finally, just as reflected waves can reinforce and reverberate, past events have given rise to new waves of human displacement and dislocation that are just beginning to swell and echo around the Pacific Rim today.

PATERNITY TEST

Edward Teller arrived in the U.S. in 1935, having left behind the anti-Semitism of his native Hungary for physics studies in Germany and, after Adolph Hitler's rise to power, Denmark. Once in the U.S., he joined the small but growing cohort of refugee scientists displaced by the political turmoil and persecution of the 1930s. Teller's involvement with nuclear weapons started early. In August 1939, Teller, along with fellow Hungarian refugee Leo Szilard, traveled to Long Island to persuade Albert Einstein (another exile from Nazi Germany) to sign a letter. The letter warned President Franklin D. Roosevelt about possible German work on atomic weaponry. It helped put in motion a chain reaction that resulted in the Manhattan Project, the full-blown Allied effort to build nuclear weapons that would be used in August 1945 to destroy Hiroshima and Nagasaki.

Whereas the engineers, scientists, and technicians of the Manhattan Project worked feverishly to design and build fission bombs (their power comes from the splitting of uranium and plutonium), Teller decided he wanted to build something entirely different. Driven by his own inner anxieties and a desire for notoriety, he devoted his time to imagining and promoting plans for a fusion bomb. This type of weapon gets its power by creating

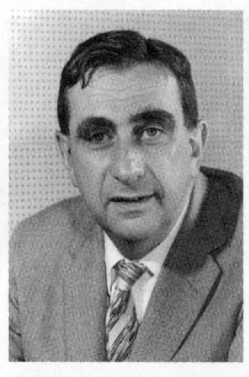

conditions similar to what happens inside a star when light elements like hydrogen combine and release staggering amounts of energy. Teller's attraction to nuclear fusion was twofold: The theoretical physics was more interesting, and thermonuclear bombs can be made much more powerful than *fission* weapons. There is no theoretical limit, in fact, to their destructiveness. Providing more nuclear fuel results in a bigger explosion.

After World War II ended, Teller's focus on thermonuclear weapons deepened, furthering his estrangement from many other physicists who found his obsessions to be unethical and his incessant lobbying for them unseemly. Phrases like "weapon of genocide" and "an evil thing considered in any light" appeared in official reports opposing Teller's vision. However, in January 1950, following the detection of the Soviet Union's first nuclear weapons test, President Harry S. Truman announced that the U.S. would "continue its work on all forms of atomic weapons," including on Teller's hydrogen bomb.[4]

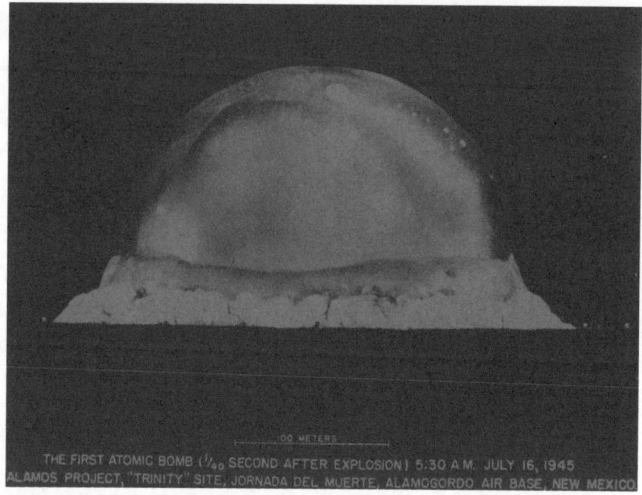

Detonation of the first atomic bomb, New Mexico, 16 July 1945. Photograph courtesy AIP Emilio Segrè Visual Archives.

There was a problem, however. Despite his relentless promotion, boasting, and lobbying, Teller had no viable design. Researchers using newly available digital computers repeatedly refuted his concepts as scientifically unworkable. As a result, Teller grew increasingly frustrated, isolated, and desperate.

Then, in March 1951, Teller and Polish-born mathematician and fellow refugee Stanislaw Ulam devised an innovative solution. It should be possible, they proposed, to direct the energy from an exploding mass of fissile material in such a way that it could ignite a thermonuclear fusion reaction. While the precise details remain highly classified, the essence of their scheme revolved around using the waves (in the form of x-rays) produced by an exploding fission bomb to compress and heat the fuel for the much more powerful fusion reaction. This new design was what Robert Oppenheimer later praised as "technically sweet."[5]

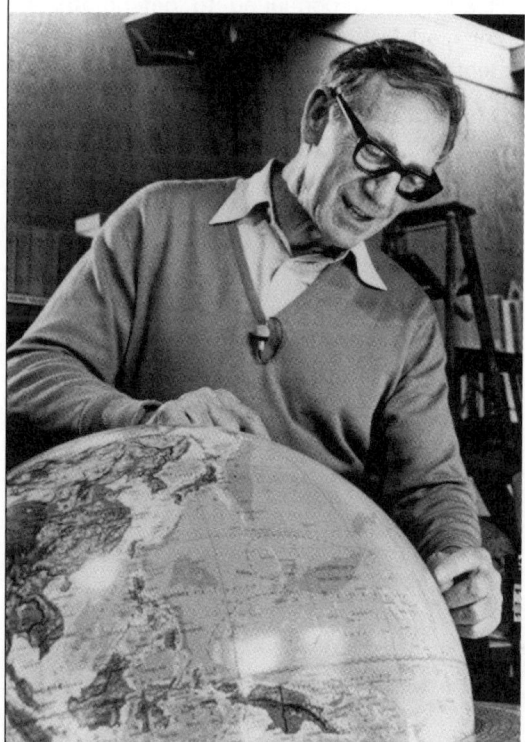

Walter Munk pointing to spots on a globe. Photograph courtesy American Geophyiscal Union (AGU), courtesy of AIP Emilio Segrè Visual Archives.

4. Peter Galison and Barton Bernstein, "In Any Light: Scientists and the Decision to Build the Superbomb, 1952–1954," Historical Studies in *the Physical and Biological Sciences* 19, no. 2 (1989): 267–47.

5. Ibid.

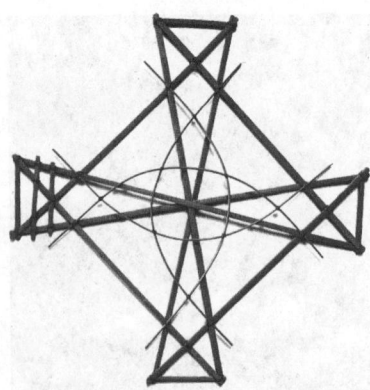

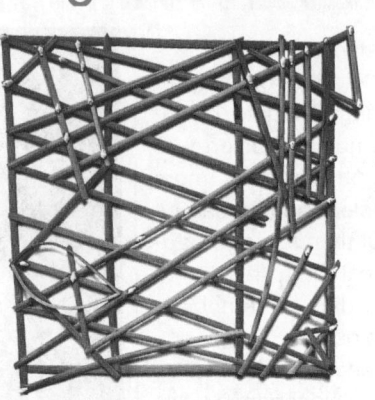

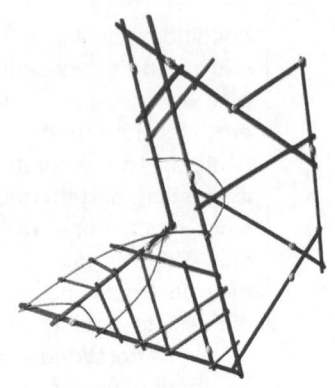

Wapepe constructed by Isocker Anwell, 2016. Also referred to as *mattang, wapepe* are Marshallese stick charts that have multiple perspectives and interpretations. This shows the navigation signs or "sea-marks" used to remotely detect an island, represented by the intersection of lines at the center of the lattice work. Photograph by Joe Genz.

Rebbelib constructed by Isocker Anwell, 2016. *Rebbelib* are Marshallese stick charts show-ing the various islands (cowry shells) of the entire archipela-go in relation to ocean swells, currents, and wave patterns. Photograph by Joe Genz.

Medo constructed by Isocker Anwell, 2016. *Medo* are Marshallese stick charts showing a particular region of the islands. Photo-graph by Joe Genz.

Theory is one thing. Determining whether the concept would work in practice was what brought Teller to Enewetak Atoll in May 1951, giving him the chance to relax briefly in the lagoon's warm waters. The test device, codenamed *George*, was placed on top of a 200-foot-tall tower. *George* exploded on May 8 with a force of 225 kilotons of TNT (one kilo-ton equals the energy released by detonating 1,000 tons of TNT).[6] Teller noted that the ex-plosion was "impressive even from my posi-tion ten miles away."[7] While the test proved the workability of the Teller-Ulam solution, Teller said nothing about the colorful local reef fish whose habitat had just been irradiated and vaporized.

It would be more than a year before a full-scale test of a thermonuclear device (as op-posed to a deliverable weapon) would be car-ried out. The *Mike* shot, part of Operation Ivy, took place on the small island of Elugelab on the northern end of Enewetak Atoll. Exploding on November 1, 1952, *Mike* produced a yield of over ten megatons of TNT, forty-five times more than *George* and 400 times that of *Trinity*, the first-ever nuclear bomb detonation. Eluge-lab was completely vaporized, replaced by a jagged crater more than a mile in diameter and 150 feet deep, while the mushroom cloud from *Mike* rose some twenty-seven miles high.

But Teller was not at Enewetak to witness the test. His relations with many of his physicist colleagues had continued to deteriorate, a point made when he was passed over as a potential leader of Los Alamos' thermonu-clear weapons program. Throughout 1952, he continued to criticize *Mike*'s design while pushing for a new nuclear weapons lab that would compete with Los Alamos (and give him a more prominent role). Bitter and angry, he sequestered himself on the day of the test in the University of California, Berkeley's Bacon Hall, where the school's geology department was located. By knowing roughly when *Mike* would explode, Teller could estimate when vibrations, transmitted as seismic waves through the earth's crust, would register on an instrument he was monitoring. His recollec-tion of seeing evidence of these waves, trav-eling to him with their burden of information, is a picture of both isolation and obsession:

> I sat down in the dark in front
> of the proper seismograph... and
> watched a luminous spot on
> a screen... About a quarter of an
> hour later, precisely when Dave
> [physicist David Griggs] told me
> it should occur, I saw the dot on
> the seismograph screen do a little

dance. The compression wave from the explosion had spent that time traveling to the coast of California… The proof was in my hand. The new approach had worked."[8]

With radioactive dust from *Mike* still filling the skies over Enewetak, Teller was eager to let sympathetic physicist friends know the test based on his idea was a success. He proudly sent them a telegram stating simply, "It's a boy."[9] In the months and years to come, Teller would become internationally known as the "father of the hydrogen bomb." It was a moniker he did little to discourage. Such is the nature of paternity. Teller would go on to design and advocate for super-weapons with yields of 1,000 megatons or more.[10] Such is the nature of megalomania. The wave that traveled to Teller across thousands of miles of ocean carried not just information but it also conveyed a sense of disasters and dislocations to come.

AN UNBROKEN SHELL

Ruprup jokur is a nautical expression in Marshallese. Translated literally as "breaking open the turtle shell," it refers to a navigational trial carried out at sea. When successfully performed, the novice navigator's mind is pictured as opening up, filling with skills and knowledge. During the *ruprup jokur* test, the student must demonstrate he (navigator status is, currently, reserved for men) can navigate by reading the movement and shape of the ocean waves. When the trial ends and the navigator has made landfall at the appointed spot, he is designated by community leaders as a *ri-meto*—a "person of the ocean." Korent Joel's path to becoming a *ri-meto* would not culminate until 2006.[11] For decades prior to this, however, his life would be affected by waves of circumstance and catastrophe put in motion by Teller and his colleagues.

During the early 1940s, Joel's family was among the few hundred people living on Rongelap, an atoll some 300 miles east of Enewetak. Bikini Atoll sits between the two locations, about 100 miles from Rongelap. Houses on Rongelap were simple constructions, usually made from palm leaves with

6. The 1945 *Trinity* test in New Mexico, by comparison, which Teller had witnessed, produced a yield of twenty-five kilotons.

7. Teller, *Memoirs*, 322.

8. Ibid., 351–52.

9. Ibid., 352.

10. The largest nuclear test to date was the *Tsar Bomba* test—some fifty megatons—done by the Soviets in 1961. See Alex Wellerstein, "An Unearthly Spectacle: The Untold Story of the World's Biggest Nuclear Bomb," *Bulletin of the Atomic Scientists*, October 20, 2021, thebulletin.org/2021/11/the-untold-story-of-the-worlds-biggest-nuclear-bomb/, accessed September 1, 2022.

11. None of the material I have presented on Korent Joel and Marshallese navigation would have been possible without the scholarship of anthropologist Joseph H. Genz. For several years, Genz carried out a series of ethnographic observations of the culture and techniques of traditional navigation skills. These are presented in Genz's remarkable book *Breaking the Shell: Voyaging from Nuclear Refugees to People of the Sea in the Marshall Islands* (Honolulu: University of Hawai'i Press, 2018). I am indebted to this book as well as a series of articles Genz and his colleagues have published.

coral and pebble floors. Their inhabitants largely lived on foods like fish, coconuts, and taro. Joel recalled that his family and their ancestors on Rongelap were once "very strong. They didn't need to look for food."[12]

After forcibly removing 167 Marshallese to a neighboring atoll, the U.S. started conducting nuclear tests at Bikini in 1946.[13]

Those two atomic explosions, part of Operation Crossroads, were almost trivial displays of power compared to what was to come. At dawn on March 1, 1954, an airplane dropped a device codenamed *SHRIMP* for a thermonuclear weapons test shot dubbed *Castle Bravo*. Based on the Teller-Ulam design, it exploded above Namu, a small island on the northwest side of the atoll. The physicists had predicted a yield of about five megatons. However, they had not accounted for a critical fusion reaction that takes place in the nuclear fuel.

To their surprise and growing horror, the *Castle Bravo* test produced an explosion three times bigger than the scientists had expected. The fireball from the fifteen-megaton blast—1,000 times more powerful than what had destroyed Nagasaki—was four and a half miles across. The searing waves of heat trapped people in observation bunkers more than twenty miles away. One physicist at the site said the discolored mushroom cloud rising over him looked "like a diseased brain up in the sky."[14]

Joel was six years old when the U.S. carried out the *Castle Bravo* test on Bikini. U.S. officials had not provided residents on nearby islands any warning of possible dangers. Standing on the roof of a house in Kwajalein, more than 200 miles away, Joel could see the bright light from the explosion and "thought it was a very big moon."[15] Later, he learned that *Castle Bravo*'s intense heat had burnt two of his grandparents.

About six hours after the test, radioactive dust began falling on the islands and atolls east of Bikini, including Rongelap. Some of it was calcium precipitated from Bikini's coral reefs as the blast had gouged a crater some 6,500 feet across and 250 feet deep. Resembling snow, the poisonous material fell across Rongelap, whitening people's hair and sticking to their clothes. Children

played in it. The dust covered coconuts on the trees. It formed a yellowish layer of water in their cisterns and catchments. The residents began feeling nauseous and their skin itched and burnt.

Fifty hours after the *Castle Bravo* test, a U.S. warship finally arrived at Rongelap to evacuate people, including members of Joel's family, to Kwajalein and other atolls. It would be three years before they could return home. Joel and his family had become nuclear refugees.

In the absence of nuclear testing, Joel could have studied traditional navigation techniques on Rongelap, learning how to become a *ri-meto* from an older navigator already experienced and recognized by the local community. Part of this would have included information from oral traditions as well as teaching devices in the form of various kinds of "stick charts." Used on land, these are not representations of islands or atolls in the ocean, as western anthropologists initially thought. Rather these tools, constructed out of shells, tree roots, and coconut fronds, serve as pedagogical instruments. They represent an idealized model of ocean swells and how these intersect, bend, and meet with one another as ocean waves move through and around the region's islands and atolls.[16]

Eventually, the *ri-meto*'s teaching would incorporate time with his apprentice on the water. A blindfolded navigator-in-training might be placed in a canoe, for example, and moved about a shallow lagoon by his teacher so as to feel the subtle variations of swell, trough, and crest. What was experienced and sensed bodily on water would, ideally, translate into a reckoning of where one was. The waves would form a path to knowing, containing information about where one was.

But, Joel noted, "the bomb stopped everything."[17] Besides the health effects due to radiation exposure, testing disrupted traditional patterns of knowledge transmission and learning. Indigenous techniques for making traditional canoes were lost. And without the *ri-metos*, there was no one to train new navigators or to administer the carefully regulated procedure that would allow one to take the *ruprup jokur* test. Marshallese wave piloting is deeply rooted in tacit knowledge and experience. The waves offer a way of knowing. They contain information but only for those attuned to their frequency and amplitude. Joel's dream was suspended in time.

SOUNDING THE CLIMATE

Walter Munk's mother, observing the rising political unrest in their native Austria, sent her teenage son into exile in the U.S. in 1932. Munk eventually made his way to Southern California, where his family later joined him. He took a job at what is now the Scripps Institution of Oceanography in La Jolla while pursuing a graduate degree from the University of California, Los Angeles. Like Joel, he was eager to study and know the ocean's waves.

On his first research voyage at sea, Munk collected measurements to look for correlations between variations in ocean characteristics as a function of the waves moving about underneath the ocean surface. He later named his modernist seaside home Seiche, an oceanographic term for a standing wave in an enclosed body of water.

During World War II, Munk enlisted in the American military as an oceanographer, a scientist in uniform. One of his first assignments was to estimate the height of breaking waves on the North African coast in advance of a major Allied amphibious landing. If the waves were too high, they would swamp landing craft and drown soldiers. The predictive method Munk helped develop was used to plan amphibious landings in Sicily, Normandy, and the Pacific. The research helped save many lives and sped up the war's end.

Munk traveled to the Marshall Islands in 1946 for Operation Crossroads and again in 1952 for Operation Ivy's *Mike* shot. For the latter, Munk's team had been assigned to measure air and water pressure waves before and after the blast at several sites. One concern the scientists voiced was that the tremendous explosion might trigger a tsunami which would submerge nearby islands and atolls. Given that the average elevation of the Marshall Islands is only six feet, with some surface features at sea level, this was a real concern. For the test, Munk stood on a raft, ready to signal semaphore codes such as *ABLE ABLE ABLE* (for a destructive tidal wave) or *DOG DOG DOG* (all clear). Floating

miles away from the test site, the wave expert still wore reflective goggles. "My memory is faulty after fifty years," he recalled, "but I will not forget the boiling sky overhead."[18] Fortunately, *DOG DOG DOG* was the order of the day.

Waves and vibrations from nuclear tests were devastating but irregular phenomena. During the course of the Cold War, however, U.S. and Soviet scientists wired the oceans to constantly monitor for potentially hostile activity via sound and vibrations. As

12. Joel, in Genz, *Breaking the Shell*, 92.

13. There is a considerable body of literature on nuclear testing in the Marshall Islands. My information on these tests, especially *Castle Bravo*, draws on two recent books: Martha Smith-Norris, *Domination and Resistance: The United States and the Marshall Islands during the Cold War* (Honolulu: University of Hawai'i Press, 2016) as well as Walter Pincus, *Blown to Hell: America's Deadly Betrayal of the Marshall Islanders* (New York: Diversion Books, 2021).

14. Richard Rhodes, *Dark Sun: The Making of the Hydrogen Bomb* (New York: Simon and Schuster, 1995), 541–42.

15. Genz, *Breaking the Shell*, 92.

16. Genz, et al., "Wave Navigation in the Marshall Islands: Comparing Indigenous and Western Scientific Knowledge of the Ocean," *Oceanography* 22, no. 2 (2009): 234–45.

17. Joel, in Genz, *Breaking the Shell*, 93.

18. Walter Munk and Deborah Day, "IVY-MIKE," *Oceanography* 17, no. 2 (June 2004): 96–105.

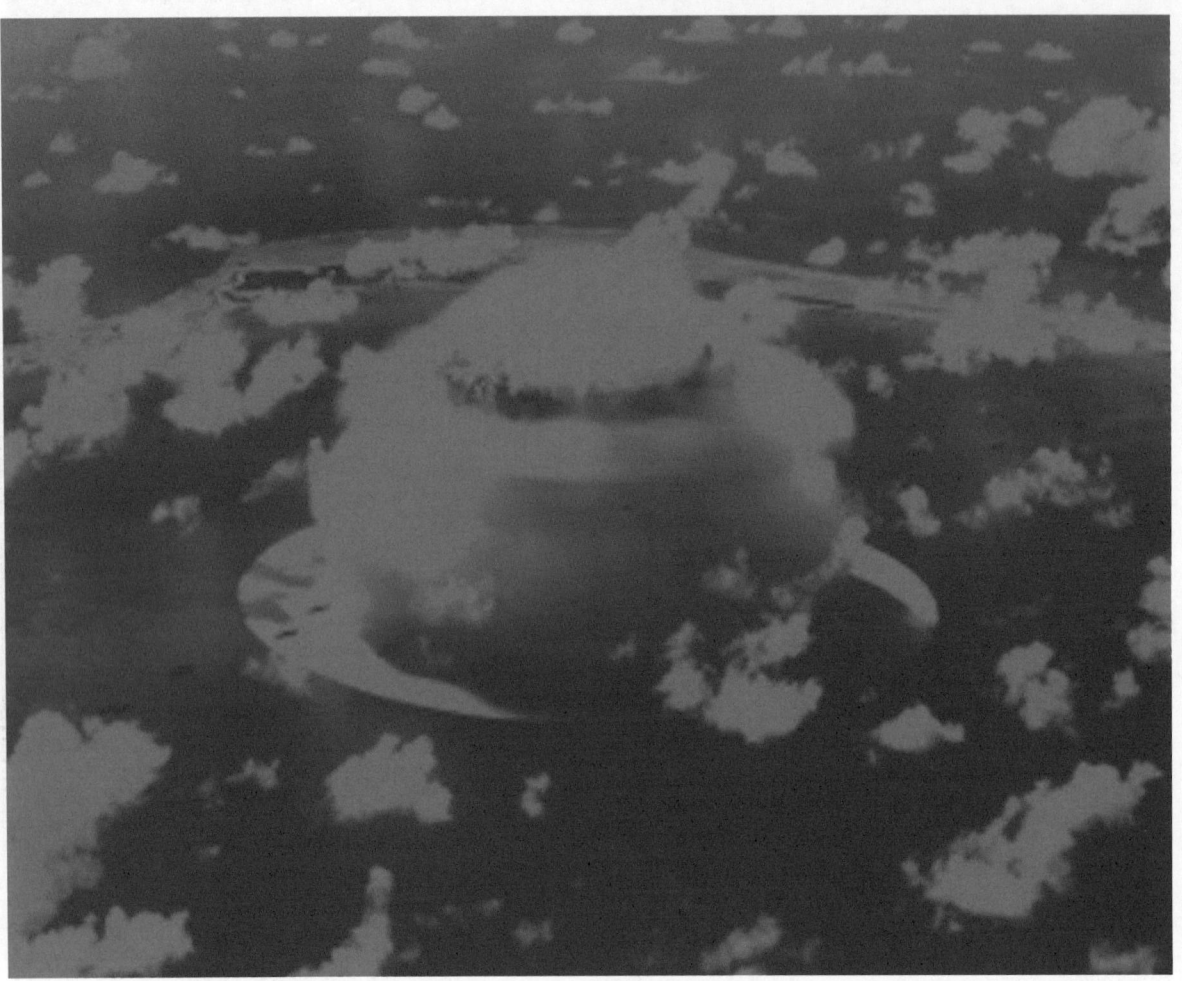

Operation Crossroads nuclear weapons test, Bikini Atoll, Marshall Islands, 25 July 1946. Photograph from the Digital Photo Archive, Department of Energy (DOE), courtesy AIP Emilio Segrè Visual Archives.

early as the *Trinity* test in July 1945, seismologists understood that the same equipment used to sense earthquakes could be deployed to detect nuclear tests.[19]

A critical issue, however, was being able to accurately distinguish between the two events. The U.S. invested hundreds of millions of dollars to build a network of stations and instruments that could detect signals from possible nuclear tests that the Soviets might conduct in violation of the 1963 test-ban treaty.[20] It was a vibrational solution to a political problem.

Another threat both the Soviets and Americans feared was the ability of submarines carrying nuclear-armed missiles to move about undetected. In the U.S., the navy generously supported research and development,

some of it done at Scripps where Munk was based, which resulted in the highly classified Sound Surveillance System (SOSUS). Hydrophones and other devices were deployed throughout the planet's water to detect and monitor the movement of Soviet submarines. Starting with a test array in the Caribbean, by the late 1950s engineers had expanded the network into the Pacific with stations operating at Midway Island and Guam. During World War II, Munk himself had done research related to anti-submarine warfare, and his familiarity with ocean acoustics led to him joining the SOSUS research group. Sound could be a way of finding as well as knowing.

Studying waves created an essential part of Munk's career. For years, he led a different project that tracked the movement of ocean swells across open water. Instruments recorded waves originating in stormy waters near Antarctica as they traveled north to finally expend their fury on the Alaskan shore. This research was done, in part, by using Scripps' new Floating Instrument Platform (FLIP). As the acronym hints at, FLIP could do just that—rotate ninety degrees from the horizontal to the vertical by filling ballast tanks in its stern. With most of FLIP's 350-foot length underwater, it was a remarkably stable platform. FLIP did not start out with a purely scientific agenda. The navy funded its construction as part of a program to build a new submarine-launched missile, armed with a 250-kiloton nuclear warhead, to destroy Soviet submarines. In 1967, a group of documentary movie makers captured the research done by Munk and his Scripps colleagues with FLIP as they studied the movement of ocean swells, "mixed and piled on top of one another in lovely confusion," for a film titled *Waves Across the Pacific*.[21]

Munk's Cold War–fueled work on submarine detection using sound waves primed him to think about what else could be learned using auditory vibrations. Across all scales, the ocean is a heterogeneous place, changing dramatically from the local to the global. The motion and behavior of waves within a small lagoon varies quite differently compared to massive swells and fetch formed across thousands of miles of open water. Turn your perspective ninety degrees and it is the same. The behavior of waves on the ocean's surface differs from the circulatory motions existing below as the epipelagic zone gives way to the mesopelagic and, eventually, the great abyssal oceanic trenches. Even when visible light is no longer detectable, these regions can still be explored and known by sound and vibration. Munk's research consistently exploited this feature.

The oceans, of course, also play an essential role in storing both heat and greenhouse gasses. But measuring the average temperature of a large swath of ocean is a challenging task, both phenomenologically (local variations obscure larger patterns of average temperature) and logistically (a few ships taking random measurements would not produce an accurate picture). Oceanographers had long known, however, that the

speed that sound travels in the ocean increases predictably as the temperature of the water increases. By measuring the speed of low-frequency sound waves, which travel more efficiently over long distances, the ocean could, Munk argued, provide a sort of global thermometer that used sound waves to make measurements.

In the late 1970s, Munk and his colleague Carl Wunsch developed a technique known as "ocean acoustic tomography." It uses sound waves as a probe to create a three-dimensional picture, over distance and depth, of the ocean's conditions over large distances.[22] By taking measurements across the planet, ocean acoustic tomography offers a way to estimate the ocean's average temperature at a particular point in time.

19. B. Gutenberg, "Interpretation of Records Obtained from the New Mexico Atomic Bomb Test, July 16, 1945," *Bulletin of the Seismological Society of America* 36, no. 4 (1946): 327–30.

20. Nilo Lindgren, "Earthquake or Explosion: The Science of Nuclear Test Detection," *IEEE Spectrum* 3, no. 8 (August 1966): 66–75. See also Axel Volmar, "Listening to the Cold War: The Nuclear Test Ban Negotiations, Seismology, and Psychoacoustics, 1958–1963," *Osiris* 28, no. 1 (January 2013): 80–102.

21. *Waves across the Pacific*, directed by Robert E. Dierbeck and Harry Muheim, McGraw Hill Text-Films, 1967, 31:42.

22. Munk and Carl Wunsch, "Ocean Acoustic Tomography: A Scheme for Large-Scale Monitoring," *Deep-Sea Research Part A. Oceanographic Research Papers* 26, no. 2 (February 1979): 123–61; Robert C. Spindel and Peter F. Worcester, "Ocean Acoustic Tomography," *Scientific American* 263, no. 4 (October 1990): 94–99.

Done successively over a long enough span of time, it would offer researchers a new and independent assessment of whether the overall temperature of the planet was changing and by how much. In 1991, Munk joined an expedition to broadcast acoustic signals from waters near Heard Island, an uninhabited and remote volcanic island some 3,000 miles off of Australia's western coast. A series of fifteen other participating ships, located at sites all around the world, detected the signals from Heard Island before a powerful gale brought the experiment to an end.[23]

After Munk's death in 2019, at age 101, colleagues recalled that the results from the Heard Island Feasibility Test was the highlight of the oceanographer's long career.[24] In a speech Munk gave upon receiving the Crafoord Prize in Geosciences 2010 in Sweden for his oceanographic research, he drew attention to the increased rate of sea rise that was accompanying global warming. Parts of the world might slip beneath the waves. Waves and vibrations transmitted across vast oceanic distances, he explained, offered scientists a method to listen, as Munk said, to "the sound of climate change."[25] It was a sound that residents throughout the Marshall Islands were already particularly attuned to.

JOURNEYS

The 1954 *Castle Bravo* test sent shock waves of disruption through the lives of Joel and his family members. Displaced from Rongelap, they were exiled from their homes for months. Indigenous peoples in areas on or near nuclear tests sites around the world were subject to similar experiences. Once the Rongelapese were back on their home islands, their cancer rates, birth defects, and other reproductive issues started to rise even as the American scientists studying them were not authorized to offer them treatments.

A few years after Joel returned to Rongelap, he began to slowly and clandestinely learn traditional navigation from his grandfather. Sometimes this would take place when they sailed from Rongelap to go fishing. On other occasions, the boy and his elder would stop at the training lagoon when the tide was low, and the young man would get some sense of the wave patterns as fluid

signposts for navigating. But, after about five years, Joel's grandfather began to develop radiation sickness, and their training sessions stopped.

Despite government claims that the area was safe for habitation, Joel left Rongelap again when he was sixteen.[26] But he remained with the waves. After finishing high school in Hawai'i, Joel studied navigation, done with compass and sextant, at a maritime academy there. He eventually earned a commercial captain's license and headed up ships transporting cargo and supplies for the Marshallese government. Throughout his career piloting large, modern ships. Joel continued his observations of the sea, connecting his own physical sensation of the waves with what his instruments told him about his place and position on the water.

In 2005, Joel met Joseph Genz, a graduate student in anthropology at the University of Hawai'i who was studying traditional Marshallese navigation and voyaging. A partnership emerged between Genz, Joel, and Alson Kelen, a younger extended family member of Joel's. Their ethnographic collaboration was remarkable for several reasons. It was an opportunity for Joel to complete the traditional navigational studies he started decades earlier with his grandfather and, with the blessing of the community *iroij* (chiefs), become recognized as a *ri-meto*, perhaps the last one still living.

The group was also interested in placing two different systems of knowledge into conversation with one another. As a boy, Joel had learned traditional Marshallese ways of extracting knowledge from the waves. Then, as a professional navigator and ship captain, he had learned and relied on modern navigational techniques and technology. Now, approaching sixty, Joel saw a chance to see if Western ways of knowing could inform, perhaps legitimate, Marshallese navigational knowledge.[27] There was great interest, for example, in what Joel referred to as the *di-lep*, a Marshallese term for a particular wave pattern. Translated as "backbone," it refers to a straight line that intersecting waves can form between two islands. According to Joel, navigators who can detect it use the *di-lep* as a guide for determining location and route finding. In time, an extended research team coalesced. Its members compared wave buoy measurements, wave model simulations, and satellite imagery with local descriptions of wave patterns and navigational concepts.[28]

Finally, the collaboration could revitalize traditional Marshallese skills associated with canoe building, sailing, and navigating. In 1996, Kelen had helped found *Waan Aelõñ in Majel* (Canoes of the Marshall Islands, or WAM) which began documenting traditional canoe design and construction techniques. In time, WAM added training for Marshallese youth and public outreach through races and displays to its mission. As part of the collaboration, Kelen apprenticed himself to the elderly Joel to learn more about indigenous navigational skills.

In early September 2006, Joel, Kelen, Genz, and a small crew left Kwajalein on the *Mali*, a thirty-five-foot sloop, for Ujae Atoll, some 140 miles to the east.[29] The voyage would provide Joel, who was navigating the *Mali* by intuition and experience, the opportunity to belatedly pass his *ruprup jokur*, the traditional navigational test. The initial conditions were not auspicious. The recent passage of Typhoon Ioke had left the waves outside the lagoon's protection a jumbled and chaotic mess.

As Genz's 2018 book *Breaking the Shell* records, it was a dramatic trip. At one point, as the *Mali* and her crew experienced rough seas and overcast skies, Genz worried that Joel would fail the test—candidates were afforded only one chance—and his opportunity to obtain community status as a traditional navigator would be forever lost. Even more threatening was the possibility that Joel would become *wiwijet*; that is, he might enter the state of deep confusion and panic that sets in when one has totally lost their sense of place and position. Despite some difficult and confusing moments, Joel navigated *Mali* safely to Ujae. The conditions for the return trip were more favorable and, on the return, Genz observed Joel's growing sense of confidence as he identified currents, swells, and wave patterns and guided their boat back to their destination at Kwajalein's southern tip. Marshallese chiefs quickly confirmed Joel's accomplishment—the shell had been broken.

THE BREAKING WAVES

There is a precise point and a place where the lives of all three of these once-refugees—Edward Teller, Korent Joel, and Walter Munk—come together to refract, combine, and interfere. It is a messy and perilous conjunction.

Runit Island is on the east side of Enewetak Island, less than ten miles away from where the *George* test was conducted

23. Ocean acoustic tomography proved controversial in the 1990s, when marine biologists and environmentalists protested against its deployment, saying that the sounds would harm marine mammals. Naomi Oreskes, "Changing the Mission: From the Cold War to Climate Change," in *Science and Technology in the Global Cold War*, John Krige and Oreskes, eds. (Cambridge, Massachusetts: The MIT Press, 2014): 141–87.

24. Chris Garrett and Carl Wunsch, "Walter Heinrich Munk. 19 October 1917–8 February 2019," *Biographical Memoirs of Fellows of the Royal Society* 69 (2020): 393–424.

25. Munk, "The Sound of Climate Change," Tellus A: *Dynamic Meteorology and Oceanography* 63, no. 2 (March 2011): 190–97.

26. The biographical material comes from interviews and other material Genz collected from Joel and appears in Genz's *Breaking the Shell*.

27. Genz, *Breaking the Shell*, 115–17.

28. Genz, et al., "Wave Navigation in the Marshall Islands," 234–45.

29. My telling of *Mali*'s journey is adapted from and indebted to Genz, "Breaking the Shell," in *Breaking the Shell*, 153–70.

in 1951. Somewhere nearby, if it was not destroyed, one might perhaps find the lagoon that Teller snorkeled in, undisturbed. But that is not likely. Forty-three nuclear tests were carried out throughout Enewetak, with the total equivalent of some thirty megatons of dynamite. Radioactive contamination is endemic throughout the islands of Enewetak. Most of the islands are completely deforested, with no coconut or breadfruit trees remaining. Crumbling concrete structures and jumbled heaps of discarded construction machinery and military equipment, some of it radioactive, is strewn over the barren landscape. At least one biological warfare test was conducted on the islands at some point.

In 1973, the U.S. government announced that exiled Enewetakese could finally return to their homes. But only after the atoll had been cleaned up. So, between 1977 and 1980, some 4,000 military personnel—many not wearing much in the way of protective gear—shoveled and scraped tons of radioactive dirt and debris into a pile. A 1958 test, codenamed *Cactus*, carried out on Runit Island, had left a sizable crater and, given the already-high level of contamination nearby, the U.S. Department of Defense decided this hole would serve as a suitable repository for the radioactive material. The workers bulldozed some 3.1 million cubic feet of toxic material into the unlined crater. They then capped it with a massive concrete dome. The lidded structure, visible in satellite imagery, sits on top of porous soil. Today, as waves break against it, seawater passes in and out of the repository, gradually releasing more radioactive material into the lagoon. There is no fence of any significance that keeps people out or signs to warn future visitors away.

Runit Dome happens to sit roughly at sea level. So, as ocean levels rise in the Marshall Islands due to the melting of glaciers thousands of miles away, the Pacific's waves are advancing further up the sides of the dome. The erosion and wave action heightens the risk that the structure will disastrously collapse and release the radioactive material entombed inside.[30]

While this itself would be a disaster, the reality is that the waters and reefs surrounding Enewetak Atoll are irreparably harmed. Of greater consequence is the fact that the Marshall Islands themselves are gradually disappearing due to the forces of climate change that Munk and hundreds of other geophysicists pointed to for decades. Any proposed temporary remedies to the slowly rising sea levels are just that. In the years and decades to come, the Marshallese may again find themselves, as Joel once was, refugees due to forces and activities outside their agency.[31] First displaced by nuclear testing, climate change and rising sea levels may create a second, greater, and permanent displacement of the Marshallese.

Waves provided Teller, Munk, and Joel with ways of knowing things—some local, some global—about the ocean and what was happening in it. Reverberating back and forth in time and space, the vibrations revealed events that had passed, provided a sense of direction, and served as harbingers of things to come. In the future, the ocean's waves will no longer just be a way of knowing. The waves will form part of our emotional architecture as parts of our world gently, deliberately, and disastrously slip beneath them. Waves and vibrations are telling us things. We should listen. <

30. Environmental journalist Susanne Rust wrote a series of articles in 2019 for the *Los Angeles Times* about the impact of nuclear testing and waste on the Marshall Islands; a multimedia version of the series was later published online. Rust, "How the U.S. Betrayed the Marshall Islands, Betraying the Next Nuclear Disaster," *Los Angeles Times*, November 10, 2019, latimes.com/projects/marshall-islands-nuclear-testing-sea-level-rise/.

31. Coral Davenport, "The Marshall Islands Are Disappearing," *The New York Times*, December 2, 2015, nytimes.com/interative/2015/12/02/world/The-Marshall-Islands-Are-Disappearing.html.

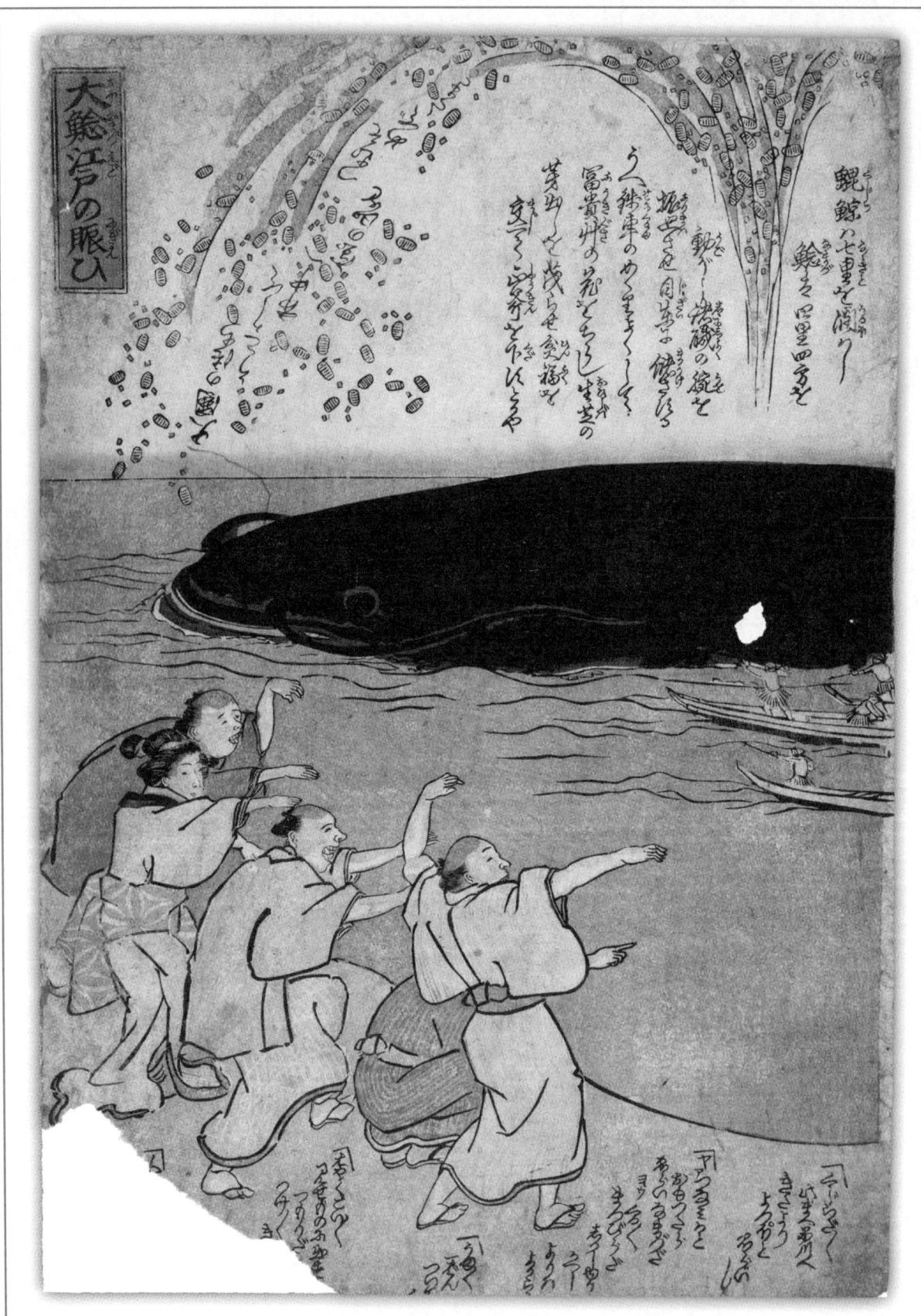

"THE OTHER RING OF FIRE"

by **ENRIQUE RIVERA**

Japanese woodblock print depicting people attempting to capture Namazu, the giant cat-fish that was believed to cause earthquakes, c. 1855/56. University of Columbia Library—Rare Books and Special Collections.

Chuang Tzu dreamed
he was a butterfly.
When he woke up
he did not know if it was Tzu
who had dreamed
that he was a butterfly
or if he was a butterfly
and was dreaming
that he was Tzu.

—Zhuangzi, c. fourth century B.C.E.

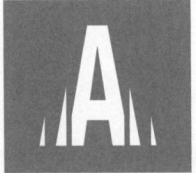 mysterious electromagnetic layer surrounds the planet, adding another physical force to our existence and creating an intangible and omnipresent electronic landscape we must navigate. We now live in a liminal geography, between the physical and digital, where our fears, desires, and anxieties configure unexpected movements. What socio-natural disasters affect this emergent and unsteady environment? And how do their uncontrollable forces manifest? These questions guide this text and test for how the irrational and oneiric might operate as referents to trace the possible tectonic faults of this hybrid reality.

TRENTREN VILU, CAICAI VILU, AND NAMAZU

We know of the existence of the Ring of Fire between North America and Asia, a 25,000-mile-long network of active volcanoes and tectonic plates from which massive tremors, earthquakes, and tsunamis emerge, causing a constant geographical reconfiguration of the planet. This activity also produces socio-natural disasters that have affected thousands of people, destroying cities and devastating economies and cultures. Aboriginal wisdom and modern science, each approaching from a different path, have narratives that explain the origin of this geography and describe its unpredictable forces.

One of the early mentions of the Ring of Fire appeared in *Scientific American* in 1878.[1] It was used to describe an area of intense

1. "The Ring of Fire, and the Volcanic Peaks of the West Coast of the United States," *Scientific American* 39, no. 2 (July 13, 1878): 26.

An illustration of the legend of Trentren and Caicai inspired by pre-Columbian designs of South-Central Chile and the South Andean area. Photograph by Dave Gelden.

seismic and volcanic activity in the Pacific Ocean, extending from the western coasts of South America, Central America, and North America, across the Bering Strait, and around to Japan and the Pacific Islands. The impact of the Ring of Fire is enormous; in Chile alone, nearly 6,900 earthquakes were detected in 2023, of which 321 were felt by the country's inhabitants.[2] This regular experience of vibrating earth has shaped Chilean culture for thousands of years. The Mapuche of southern Chile and southwestern Argentina keep a knowledge transmitted by the generations of Indigenous peoples through oral and visual tradition.[3] They understand earthquakes as phenomena produced by the supernatural beings Trentren Vilu and Caicai Vilu, giant snakes said to live in central Chile's Araucanía region.

The sons of warring Pillan, powerful nature spirits, Trentren and Caicai were turned into snakes as punishment. Trentren is a female serpent and protector of the earth, while Caicai is a male sea serpent and protector of the sea. As the story goes, one day, Caicai woke from a long sleep and, upon seeing humanity's ingratitude for all the sea had given them, he became enraged and used his fish-shaped tail to hit the water. This generated a great tidal wave that flooded the valleys and hills and carried inhabitants out to sea. Trentren, seeing the desperation of the tidal wave's victims, took the drowning people and animals on her back and delivered them to the highest and safest places. She also ordered the hills to increase their height to resist the rising waters.[4] Caicai, angry at her interference, fought Trentren until both serpents were tired. Trentren was partially victorious because not all the land was flooded, although the waters never fully retreated. The region was calm for many years after the battle, and communities enjoyed their lives in peace. Then one day, Trentren, angered by the callousness of humanity, made the earth tremble and all the volcanoes erupt. Since then, she has continued to manifest her anger through tremors, earthquakes, and volcanic eruptions, while Caicai continues to cause tidal waves and floods. Mapuche believes the only way to keep the mutated Pillan calm is human sacrifice.

This story is an explanation for how Chile obtained its present geography: the country would have been a single strip of land joined in its entirety to the South American continent were it not for the battle of Trentren and Caicai. Through the story, the Mapuche developed practical knowledge about how to respond to earthquakes and tsunamis, including building systems for earthquake-resistant housing and structures, choosing safe locations for settlements, and passing on survival techniques. As Rodolfo Lenz describes:

> To understand how the Mapuche explain in their imagination the phenomenon of the tremor, it is necessary to mention that, according to all probabilities, the religious concept of the ancient Mapuches considered as the most powerful of all the deities the Pillan, as [Andres] Febrés… explains: "Pillan, pillan, the devil, or a superior cause, which they say makes thunder, lightnings, and explosions [reventazones] of volcanoes.[5]

2. "Grandes Terremotos en Chile," Centro Sismológico Nacional, University of Chile, csn.uchile.cl/sismologia/grandes-terremotos-en-chile/, translated by author.

3. There are several hypotheses about the origin of the Mapuche. In this context, I take as a reference the study of Grete Mostny, who traces their origin to between 500 and 600 B.C.E., following the autochthonous theory. See Mostney, *Prehistoria de Chile* (Santiago: Editorial Universitaria, 1971).

4. Julio Vicuña Cifuentes, *Mitos y supersticiones recogidos de la tradición oral chilena con referencias comparativas a los de otros paises latinos* (Santiago: Editorial Universitaria, 1915).

5. Rodolfo Lenz, *Tradiciones e ideas de los araucanos acerca de los terremotos* (Santiago: Imprenta Cervantes, 1912), 7, translated by author.

The Mapuche worldview tends toward a deep respect for nature and the earth, leading to a nuanced understanding of geological events and an aspiration toward interdependence with the world around them.

At the other end of the Ring of Fire, we find ancient wisdom from the millenary Japanese culture about the region's tectonic activity. The concept coined for this condition, *tenbatsu* (天罰覿面), refers to immediate divine punishment for wrongdoing. Japan's visual and oral representations of earthquakes have survived thanks to cultural and religious traditions used to manipulate people into believing that the occurrence of earthquakes or tsunamis are connected to individual behaviors. These allow us to establish a narrative consideration of geological forces within a cultural and historical reading of Japan's relationship to the Ring of Fire. One of the most popular stories is that of Namazu (鯰), the deep-sea Catfish. Whenever Namazu moves, he produces earthquakes and tsunamis. The only way to stop him is with a sacred rock called *kaname-ishi*, which the *kami* of Kashima, a powerful deity, holds over him to keep him still. However, if Kashima is distracted, Namazu can wriggle free, causing death and destruction as well as the rearrangement of Japan's geographical elements. Guided by this story, the Japanese have developed collective orientations for surviving the effects of the earth's unceasing transformation.

COMMON MYTHS

In many stories and texts, natural catastrophes are a means of punishing those who move away from the religious precepts.[6] Correcting cosmic imbalance requires correction, often in the form of penance, scourging, or even sacrificing loved ones. These narratives have a common sequence: bad decisions, development of conflict, catastrophic end as a warning, redemption or punishment. Today, the "economy of attention" is also used as a means of control.[7] Transnational corporations fight for your awareness on social media, converting these platforms into sacrifice zones through the exploitation of human vulnerabilities.[8] But it's not just religions and corporations that use powerful and uncontrollable forces of nature to their advantage. This tactic is also fundamental to a generation of artists who, alarmed by the current massive and unforgivable economic and climate crisis, are beginning to use scientific means to interpolate this reality.

However, it is not possible, nor perhaps is it even desirable, to establish a parallel between the intensity of an earthquake and a work of art. Aesthetic constructions that evoke the experience of earthquakes have the impossible task of viscerally communicating the sensations of fear, anguish, and even excitement provoked by the movement of the earth. In 1807, the German Romantic writer Heinrich von Kleist penned *Das Erdbeben in Chili* (The Earthquake in Chile), a short novel that represents the world without having experienced it. The European intellectuals who imagined everyday life in South America interfered with the constitution of an American identity all its own, raised from its core. Eurocentrism eliminates the endemic writing methodologies of Indigenous peoples through their brutal genocide to occupy, extract, and commercialize the natural resources of their land. From 1492, when natives of the Guanahani discovered a group of navigators

6. "The earth also was corrupt before God, and the earth was filled with violence. And God looked upon the earth, and, behold, it was corrupt; for all flesh had corrupted his way upon the earth. And God said unto Noah, 'The end of all flesh is come before me; for the earth is filled with violence through them; and, behold, I will destroy them with the earth." New King James Version, Genesis 6:11–13.

7. As Thomas H. Davenport and J. C. Beck define it, "Attention is focused mental engagement on a particular item of information. Items come into our awareness, we attend to a particular item, and then we decide whether to act." Davenport and Beck, *The Attention Economy: Understanding the New Currency of Business* (Cambridge, Massachusetts: Harvard Business Press, 2001), 20.

8. "[I]n an information-rich world, the wealth of information means a dearth of something else: a scarcity of whatever it is that information consumes. What information consumes is rather obvious: it consumes the attention of its recipients. Hence a wealth of information creates a poverty of attention and a need to allocate that attention efficiently among the overabundance of information sources that might consume it." Herbert A. Simon, "Designing Organizations for an Information Rich World," in M. Greenberger, ed., *Computers, Communications, and the Public Interest* (Baltimore: The Johns Hopkins Press, 1971), 40–41.

led by Christopher Columbus on the verge of death and rescued them, to the late eighteenth-century voyages of explorers like Alexander von Humboldt, the flow of information reaching the ears of the European court distorted first-hand observation to feed the construction of an American imaginary that included fantastic animals, humans with supernatural powers, and unimaginable riches. Today, we would call these stories "fake news," constructed to manipulate collective decision making. But at that time, they built an alternative reality, an imaginary cartography, the "invention of America," annihilating and denying the existence of millions of people who inhabited this territory in more sophisticated ways than their conquerors imagined and massacring civilizations that had built their own agencies and ways of adapting to the world for centuries.

One of the subproducts of this process was an aspect of eighteenth-century German Romanticism marked by the differentiation between the Apollonian drama of Johann Wolfgang von Goethe and the Dionysian drama of Heinrich von Kleist. It is important to understand the Spanish court was rooted in the Austro-Hungarian Empire. The Spanish emperor was of Teutonic descent, and the methodologies of Prussian soldiers are still used by Latin American armies. Kleist was influenced by the news coming from America—news not only of riches but also of earthquakes, tsunamis, and all kinds of heartbreaking narrative constructions arising from delirious imaginings about the "new continent." This created a portrait in which desires and fears were constitutive elements of a reality populated by what we understand today as "hallucinations" in artificial-intelligence slang. Constructing a psychogeography from second- and third-hand sources, Kleist used a disastrous earthquake—the disorder of the stars[9]—as a plot for a romance in which violent death is the morbid result of the tragedy of love and the monsters are obtuse religious zealots.

The royal Spanish court was the official space for news from the *virreinatos*. However, news overflowed through other channels, feeding the morbid imagination of the mobs. As Chilean historian Miguel Rojas Mix observes:

> All cultures have monsters, and their origins usually have several reasons. Some monsters are specific to a territory and their existence is rather anecdotal, and others are universal and seem to resist epoch after epoch, without respecting the changes of empires or ruling cultures. It should come as no surprise that monsters originated in these lands. Monsters are something like the golems that cultures create from their fears, xenophobia, anathemas, or the unknown, among other factors. It is a reflection of the worst of themselves, of their prejudices and contempt, of their most atavistic and deepest fears. America was the New World: a huge piece of land to be discovered. It was populated by monsters long before it was explored."[10]

Chile was part of an obscure European mythology, one that sculpted part of occidental art, and the last country on earth where violent

9. The word "disaster" comes from the Latin *dis* (unfortunate) and *astrum* (star), expressing an unfortunate phenomenon produced by fate or the gods and beyond human control.

10. Miguel Rojas Mix, "América se pobló de monstruos mucho antes de ser explorada," interview with Pablo Espinoza, *La Fuente* (2015), fundacionlafuente.cl/entrevistas/miguel-rojas-mix-america-se-poblo-de-monstruos-mucho-antes-de-ser-explorada/, translated by author.

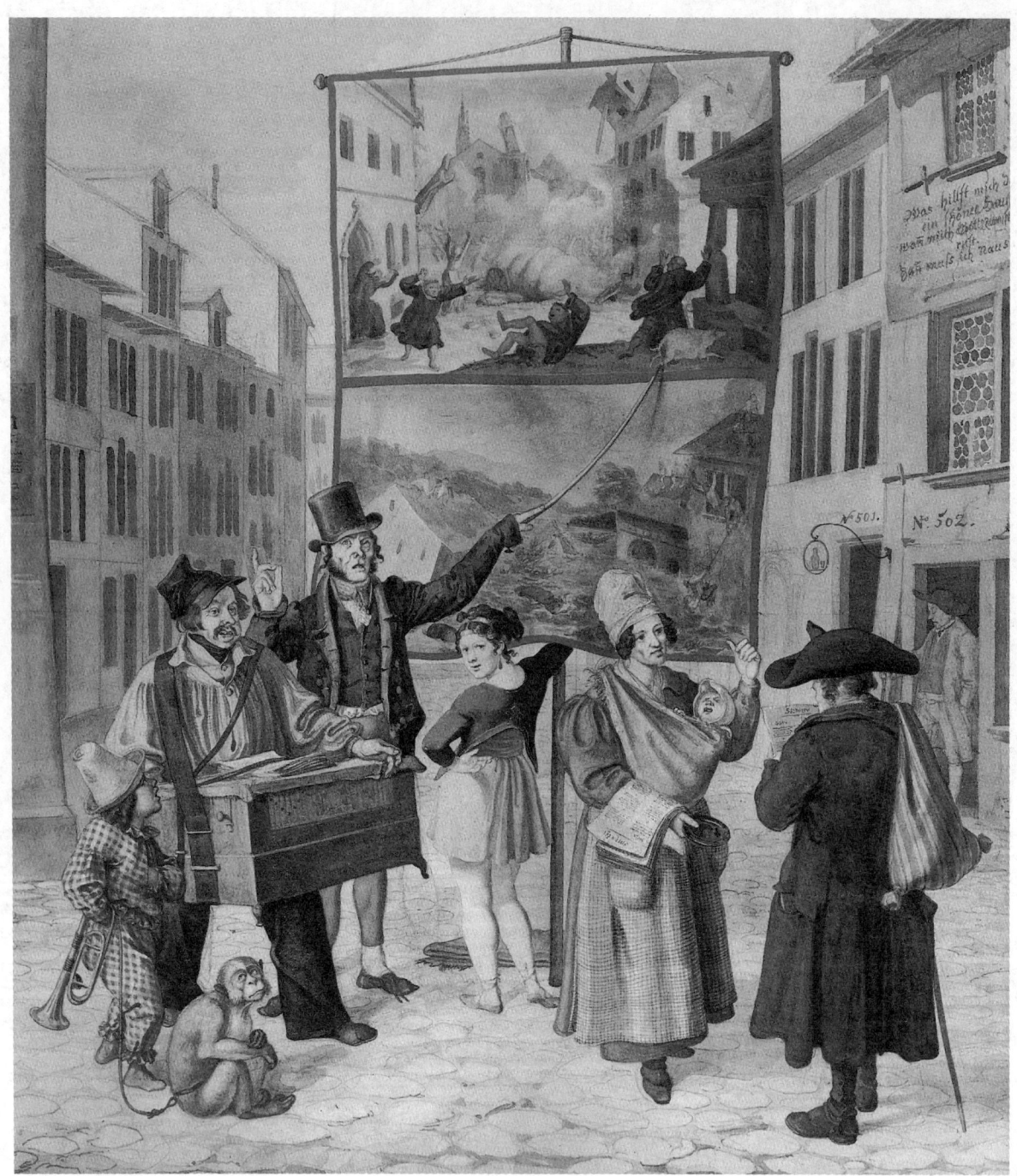

Story-singers point to images of the 1356
earthquake in Basel and the 1830 flood in
Hölstein. Painting by Hieronymus Hess,
c. 1830-50, from Eugen A. Meier, *Ein
Bildband mit Geschichten aus der Anek-
dotensammlung von Johann Jakob Uebelin
(1793–1873)* (Basel: Springer, 1970).

natural socio-disasters occurred frequently. Romanticism replaced the myth. The construction of a premodern imaginary exalted the abstract construction of a landscape close to hell, where Latin America acted as host to a scenario imagined from afar, distances that have not only been given in geographic and abstract terms. The complexity of the world, its capacities for dynamic change, but also the specter of geological augmentation over millennia coalesce in artworks such as these, opening the way for qualitative readings of these conditions we might struggle to quantify. But these are precisely the keys to understanding how we are imagining and inhabiting the internet as a continent with its own physical rules, with its own tectonic movements, with its own monsters, where art, narrative, and the collective construction of imaginaries are fundamental to compose the instruments that allow us to navigate in this electromagnetic territory that has its own laws, its own disasters, and which apparently we still do not fully understand.

BUT WHAT IS ART?

This question is formulated by Marcel Duchamp in his 1957 text "The Creative Act", which begins:

> Let us consider two important factors, the two poles of the creation of art: the artist on the one hand, and on the other the spectator who later becomes posterity. Clearly, the artist acts as a mediumistic being who, from the labyrinth beyond time and space, seeks the way out into a clearing. If we grant the artist the attributes of a medium, we must then deny him the state of consciousness on the aesthetic plane of what he is doing or why he is doing it. All his decisions in the artistic execution of the work rest on pure intuition and cannot be translated into self-analysis, spoken or written, or even thought.[11]

Malena Szlam's audiovisual portrait of the Chilean landscape depicts the driest desert in the world, one dramatically sculpted by massive earthquakes, alluviums, and the industrial extraction of minerals. Duchamp's text can be used to describe the intuitive way the work unfolds the intangible sensations the territory inspires. For *Altiplano* (2018), Szlam traveled to the desert of northern Chile to record the topological phenomena caused by the trembling of the earth—sand waves that ripple, seemingly imitating water. In the film, sound complements image with a power in and of itself. As she observed:

> I think the experience of listening to sound without images is more assimilated than seeing images without sound. It seems to be more immediate to visualize images when there is sound than to imagine sounds when there are images. Perhaps this is because moving images have historically been more closely linked to sound. In *Altiplano*, I worked with infrasound from nature, frequencies that are below the audible spectrum of the human ear: inner voices of the earth, subway water, volcanoes, and whale vocalizations.[12]

11. Marcel Duchamp, "The Creative Act" (1957), in *The Writings of Marcel Duchamp* (Boston: Da Capo Press, 1989), 138.

12. Malena Szlam, "El momento de filmar es bastante sensorial. O sucede o no," interview by Felipe Blanco, *La Fuga* (2019), lafuga.cl/malena-szlam/963, translated by author.

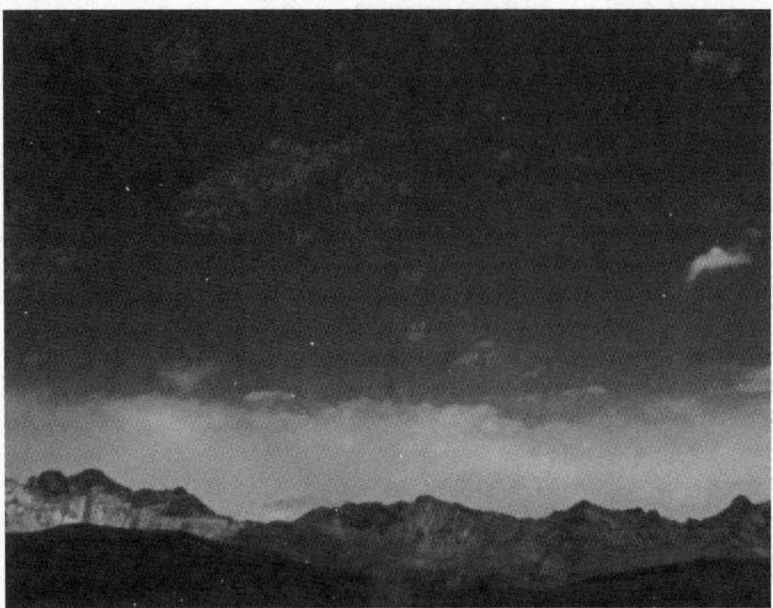

Malena Szlam, still from *Altiplano*, 2018. Photograph courtesy the artist.

13. Thierry de Duve, *Pictorial Nominalism: On Marcel Duchamp's Passage from Painting to Ready-made*, trans. Dana Polan (Minneapolis: University of Minnesota, 1991), 160.

The method is reminiscent of Duchamp's "infrathin"—everything that requires our maximum attention to assimilate what is outside our everyday perception. Like Kleist or Szlam, we approach it from perplexity, and, as Duchamp describes, "one can only give examples of it."[13] A powerful aspect of Szalm's work is observing the perpetual aftermath of an earthquake, the phantom of a disaster. In this way, Kleist and Szlam propose two different ways of accessing the sensitive layers of context that surrounds us, allowing an awareness of a wider reality that can only be revealed through traumatic events (such as an earthquake) or deep meditation provoked by profound observation, which merges with our perception and body, eliminating the borders between the work and ourselves.

ELECTROMAGNETIC PLATES

It is odd to realize that we live in a hinge time between analog and digital without a good understanding of the subtleties that exist between the two environments. Both are part of reality, yet we are still understanding what it means to share our time between the two spaces.

We are surrounded by nonlinear digital environments, giving us experiences never lived before in the history of humanity. These environments bring new fears, new traumas, new loves, new friendships, new economies, new politics, new wars, new worlds, encouraging expanded relationships with our nervous systems and maps we still do not understand. And yet we throw ourselves thoughtlessly into these environments, these digitally framed mirror worlds, without measuring the consequences and implications of their increasing density and depth.

Alba Triana's *Music on a Bound String No. 2* (2015) visualizes what happens to the environment in an earthquake, as solid becomes liquid, the atmosphere reflects colors, the air gets full of electricity, and the sound of the vibration configures a moment of earthly singularity. The work is a threshold between worlds, activating the possibility of feeling secure within the intensity of disaster as the liquefaction process generates a massive number of new illusions. The work highlights various sensitive layers, and the transformation is from one state to another, in this case, from solid, thin matter to sub-thin matter. This liquification of reality is what sociologist Zygmunt Bauman calls a liquid modernity, where all relationships undergo a process of dissolution:

> Was not modernity a process of liquefaction from the start? Was not "melting the solids" its major pastime and prime accomplishment all along? In other words, has modernity not been "fluid" since its inception? These and similar objections are well justified, and will seem more so once we recall that the famous phrase "melting the solids," when coined a century and a half ago by the authors of *The Communist Manifesto*, referred to the treatment which the self-confident and exuberant modern spirit awarded the society it found much too stagnant for its taste and much too resistant to shift and mold for its ambitions—since it was frozen in its habitual ways. If the "spirit" was "modern," it was so indeed in so far as it was determined that reality should be emancipated from the "dead hand" of its own history—and this could only be done by melting the solids (that is, by definition, dissolving whatever persists over time and is negligent of its passage or immune to its flow). That intention called in turn for the "profaning of the sacred": for disavowing and dethroning the past, and first and foremost "tradition"—to wit, the sediment and residue of the past in the present; it thereby called for the smashing of the protective armor forged of the beliefs and loyalties which allowed the solids to resist the "liquefaction."[14]

14. Zygmunt Bauman, *Liquid Modernity* (Malden, Massachusetts; and Cambridge, England: Polity Press in association with Blackwell Publishers, 2000), 2–3.

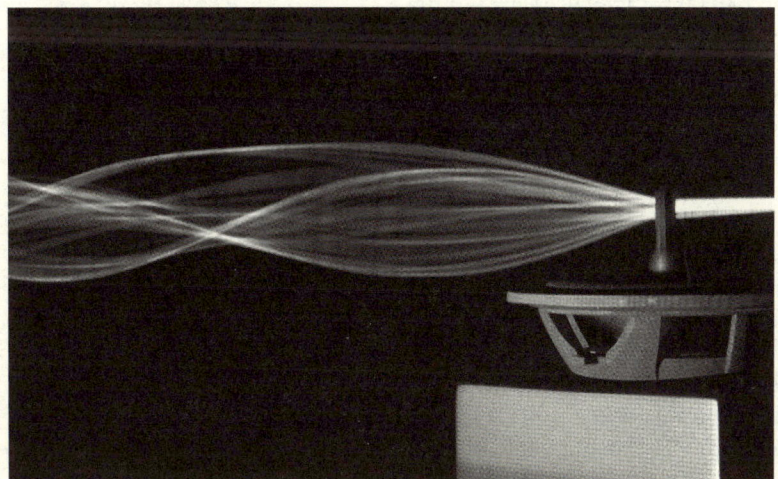

Alba Triana, *Music on a Bound String No. 2*, 2015.
Photograph courtesy Alba Triana Studio, Ernesto Monsalve.

We involuntarily confuse reality with represented reality, a condition filmmaker Alicia Vega warned poor populations of Chile about during the 1980s. Vega taught children living under dictatorship about the dangers of watching television, popular entertainment manipulated by the military regime to subjugate the collective will[15]—the precise moment of the acceleration of this liquefied reality.

We have configured an abstract map, compounded by electromagnetic impulses, algorithms, and audiovisual stimulus, that we still travel cautiously and intuitively based on a voluble language of signs. We have conceptually nurtured this territory for a long time without knowing we were doing so, from the beginnings of understanding numerical patterns, then by cybernetics, data science, complex systems, and media studies. It is a trans-generationally incubated environment configuring a territory that forces us to move without knowing its real physical, economic, and political laws, which mysteriously becomes as we move through it, like an unstable labyrinth in motion.

Gravity in this territory does not push us towards the earth. On the contrary, gravity behaves in an entropic way, privileging emotions over rational thought. As if we were moons or satellites that surround the planet, our movements configure small earthquakes that affect our environment in unpredictable ways. The manifestations that emerge from art—perhaps from unconventional, intuitive, wild art forms—are seismographs of a reality still in comprehension, observatories of a dislocated reality, in which atmospheric disturbances are captured through languages still too mysterious for our emerging expansion of sense and the capacity to perceive. German Romanticism perhaps offers a way to overcome the abstract and distant traits that emerge from this first exploration of digital environments, plagued by hatred and rage. The love and feeling of belonging and care for the natural environment might be invoked as a kind of resistance to how we are nurturing the huge learning machine that represents the internet. There is an irresistible attraction to an abstract territory emerging from electromagnetic signals, where the arts have explored the sensibilities of the materiality of this brave new electronic world thanks to ambient, drone, and electroacoustic experimentation. These are the keys to navigate the luminous side of this wild territory, plagued by simulations of the real, which emerge due to the inexperience of its early inhabitants who still strive to inhabit an untamed territory, in which the untamed is us.

A HUMAN RING OF FIRE

To understand the new laws and protocols of this mixed reality, we must be flexible, be like water, as Bruce Lee said. This new paradigm requires deconstructing our way of inhabiting the world. Our generation, perhaps before its death, will live a complete transition if it manages to deconstruct its vision of the world and begin to apply strategies that understand the chemical, physical, and electromagnetic aspects of this other dimension that is coupled in the center of our reality. Minoru Sato's work provides us with a portal to witness, also from perplexity, as in the case of Kleist and Szlam, a landscape that does not correspond to the formal territory, but to one composed of signals, vibrations, and infrathins. The work functions as a medium

15. Filmmaker Alicia Vega carried out a series of film workshops in vulnerable populations in Chile during the Augusto Pinochet dictatorship. Vega's project can be understood as one of perceptual resistance by providing a safe space for the children who attended the workshops as well as a space of resistance and containment to support their current reality and to prepare their intuition in the face of the high level of media manipulation initiated during that period. See Ignacio Agüero, dir., *Cien niños esperando un tren* (1988), cinechile.cl/pelicula/cien-ninos-esperando-un-tren/.

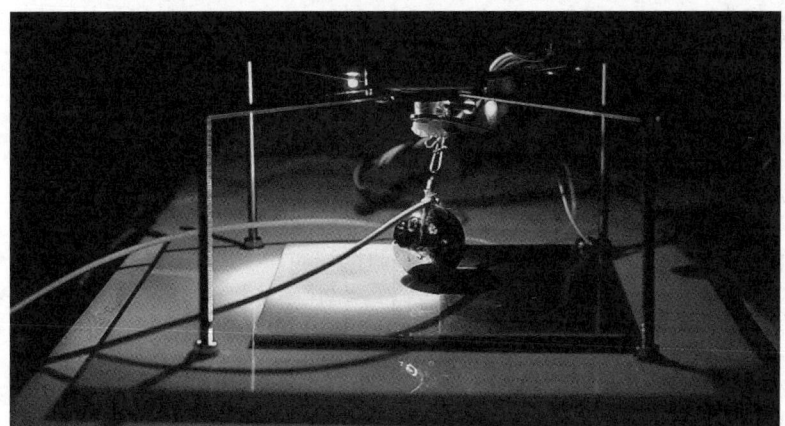

Minoru Sato, *Thermal Acoustics*, 2013 – ongoing.
Photograph courtesy the artist.

16. Abraham Cruzvillegas, "Abraham Cruzvillegas: 'La obra debe ser autónoma, se tiene que liberar al público de una imposición de significado,'" Pontifical Catholic University of Chile, 18 October 2019, artes.uc.cl/noticias/tautologia-sin-titulo-interpretacion-y-creacion-en-nueva-exposicion-de-galeria-macchina/, translated by the author.

17. Ibid.

between the visible and the invisible world, and through a series of devices and complex systems cohabit harmoniously. It makes appear what we generally do not perceive, constituting itself as a prosthesis of the intangible. It is an example of how we must adapt our perception, domesticated by hundreds of years of cultural assimilation, and understand that if this artwork exists, it was because a professional artist developed rigorous research, in which he combined concepts, aesthetics, and materials that emerge, intuitively and rationally, from an artistic production that points out routes to assimilate our mixed reality.

A word of advice: Mexican artist Abraham Cruzvillegas states that the very construction of the world, through a determined involvement with the factors of our environment, can be put into practice from a technology without hegemonic devices. We must be "as horizontal as possible, overlooking the hierarchies and the tension that institutions have with the public."[16] On the question about the relationship between the work and community, Cruzvillegas elaborates:

> In general, there is a space of tension between the work and the public, and the institution works for a didactic overpopulation, or as I would say, didacticism. For me, [the institution] almost always fulfills a very crude function, since it subjects the visitor to the idea that it is not capable of constructing content on its own. I believe that no work deserves an explanation, and that is not something I invented, but it was said by hermeneutics and it is constantly under discussion. According to me, the work must be autonomous, the public must be freed from an imposition of meaning, or from the artist's biography… I like to arrive and discover how the other appropriates what I imagine as something almost childish. What for me is an achievement can be a setback for humanity, but that makes me very happy… Self-construction… leads to a reflection on solidarity work and camaraderie when people are together, in a space of production, but which in turn is festive, warm, organic, unstable and precarious.[17]

The self-construction of meaning proposed by Cruzvillegas after a social and natural disaster is an act of survival, and both developed, developing and undeveloped countries apply logics of self-construction from their most vulnerable communities to rebuild their cities. Self-construction is also a core way to understand hacker culture, where the development of a code can be improved through the iteration of diverse programmers seeking a common good. Artistic self-construction in a context of digital analog toponymic hybridization is the basis for understanding that the creation of one's systems is the basis for not relying on hegemonically built systems (William Blake), or that if you don't program, you will be programmed (Douglas Rushkoff).

It is worthwhile to stop at this point and analyze a possible tension, and atmosphere of what we understand by art, science, and technology. Usually, the Greek notion of *techne* is established as the matrix space where these trades cohabit. However, this vision leaves out other comprehensions; diverse civilizations before the Greeks have already practiced aesthetic, technical, and philosophical production, decoding the world perceptually and sensorially in their own way. Perhaps none of these notions are valid for composing in this hybrid analog-digital continent, and it is necessary to incorporate new parameters based on the physical rules of this space. For example, there is no gravity on the internet, and the insistent tendency to create digital twins to imitate museum spaces seems more like a poor attempt to give an aura to the exhibition space in digital environments. Projects like Sphere in Las Vegas or the trend of transforming paintings by Vincent van Gogh or Claude Monet into immersive digital installations become havens for sensational experiences that stimulate the nervous system via a dopaminic spectacle. It is hard not to associate Sphere with Plato's cave, to think how thousands of people gather not in front of a screen but inside it, forgetting for a few hours the terror of the world and surrendering their nervous systems to a spectacular immersive experience. Are these the spaces that the media arts should conquer? Aren't these infrastructures the ones that should be the homes of aesthetic representations based on electric and electromagnetic media? Karlheinz Stockhausen's dome for the 1970 World Expo in Osaka, Japan, is a reference that places us in the middle of this invocation.[18]

There is a dense history of artists who have made works based on electronic and electromagnetic media, inhabitants and explorers who have intuited the confirmation of digital spaces of cultural representation. What are the new narratives that emerged from their imaginaries? What are the media aesthetics of this electronic new world? We need to search the futuristic visions of William Gibson or Bruce Sterling, the cyberpunk imagination of Ridley Scott, even the dystopia of the Wachowski Brothers or the nostalgic aesthetic of Andrei Tarkovsky. We need to examine the audiovisual explorations of Juan Downey, Nam June Paik, Steina and Woody Vasulka, and Bill Viola, the corporal transformations of Stelarc, the performative acts of E.A.T., Carmen Beuchat, Enrique Castro-Cid, Waldemar Cordeiro, and Silvia Palacios, and the visual-philosophical explorations of Yuk Hui, Ronald Kay, and Friedrich Kittler. These artists and philosophers are among

18. Germany built the world's first spherical concert hall for the 1970 World Expo in Osaka, Japan. It was based on art concepts by Karlheinz Stockhausen and an audio-technical plan from the electronic studio at the Technical University of Berlin. The audience sat on a sound-permeable grid just below the center of the sphere, while fifty groups of loudspeakers arranged around them reproduced, fully in three dimensions, electro-acoustic sound compositions specially commissioned or adapted for the space. Works by composers including Bernd Alois Zimmermann and Boris Blacher were played from the multitrack tape along with those by Johann Sebastian Bach and Ludwig von Beethoven. Over the course of the 180-day exhibition, Stockhausen and a nineteen-piece orchestral ensemble gave live concerts for over a million visitors; *Spiral*, for a soloist and short-wave receiver, was played over 1,300 times, for example. It was possible to achieve the three-dimensional sound distribution live using a spherical sensor built in Berlin to feed the fifty speakers, but a ten-channel rotary mill constructed to Stockhausen's design was deployed more frequently. See Golo Föllmer, "Karlheinz Stockhausen 'Spherical Concert Hall,'" Media Art Net, medienkunstnetz.de/works/stockhausen-im-kugelauditorium/.

the many who, through early use of electronic and computational media between the 1960s and 70s, posed a transversal critique of the use of socio-technologies, already prevented by movements such as Fluxus and Surrealism or exalted by Futurism, by declaring an unrestricted love for speed and the machine over the imperfection of human nature. These artists were warning us about how techno-solutionism acts more like the mirage of a nonexistent oasis.

We might consider "anarchitecture," a neologism coined by Gordon Matta-Clark, as a method to transform already existent infrastructures, changing their meanings with cuts that deconstruct and resignify with infrathin operations, like the way hackers, through simple and precise incisions, transform a system. This is a valuable metaphor to understand a possible regeneration of the world through the overcoming of these other earthquakes. The social role of art, science, and technology, the possibility of subjugating any hint of visual overstimulation and provoking an encounter between people who have poured their own experiences and feelings into this context can give us the parameters for an alternative narrative that assimilates the earthquakes of this other Ring of Fire. <

HOW THE FISHBOWL CHANGES

A COSMIC VIEW FROM THE INSIDE

In 2022, NASA's James Webb Space Telescope released the deepest and sharpest infrared image of the distant universe. Webb's First Deep Field shows many overlapping objects at various distances, beyond the red edge of the rainbow. Photograph courtesy NASA/ESA/CSA/STScI.

by RANA X. ADHIKARI +
AKANKSHA TIWARY

SPACE: NEITHER STATIC, EMPTY, NOR DARK

1. Scientists used to date the big bang at fourteen billion years ago, but lately there's been some controversy, so stay tuned!

f you want to understand the universe, you need to understand relativity. It feels like a heavy lift, but it's so satisfying, in the end.

The night sky is filled with stars, but they're not just twinkling. Stars, galaxies, and clusters of galaxies are all moving away from us, and the ones that are farther away are speeding away even faster. This is what we know so far: between 12 and 20 billion years ago,[1] there was a really big bang, and our universe is the remnant of that explosion. In what would seem like an instant to us, the universe expanded and cooled. Then it slowly stretched out over billions of years, giving birth to the first stars and galaxies. And it has been expanding ever since, like a blueberry muffin that never stops rising (in this metaphor, all the stars are blueberries). While space might seem like a vast emptiness, it's teeming with activity and substance.

The universe is vast and always dynamic. One of the most revolutionary discoveries in cosmology was the observation of the speeds of distant stars, as supported by the observation of the colors of distant galaxies, leading to the formulation of the big bang theory. The precise measurement of these colors led to the discovery. Just as our ears detect the pitch of sirens chirping up if they are coming towards us, or chirping down if they are moving away, our eyes and brains interpret the frequency of light waves through shifts in color. If the stars are moving away from us, they move a little toward the red side of the rainbow: blue becomes green, yellow becomes orange, and red disappears into the invisible infrared.

But what does it mean when scientists say that the universe is expanding?

It's not that the moon is going away—the earth and the moon are bound by gravity. Gravity is also not going to pull the earth away from the sun, thereby cooling the planet. Rather, the expanding universe means that the distances between galaxies are increasing over time. It tells us that the universe is not static; space itself is expanding, and everything in the universe is moving with it. There is no such thing as a truly static object or location in space. In a rising blueberry muffin, each blueberry would see all the other berries moving away. In a blueberry muffin expanding at near light speeds, each blueberry would think it was surrounded by an array of colored berries: the closest ones would be turquoise, their neighbors might be chartreuse, and the farthest away would be a fierce crimson.

What all of Albert Einstein's equations and the astronomical observations that followed seem to imply is that empty space isn't really that empty. Time and space are interwoven and inseparable, like matter and energy, mind and body, art and science. And if you put a lot of heavy stuff into a tiny space, it can have enough gravity that it bends space and, therefore, light. If you bend light enough, it just all falls into a tiny point. Once the light gets trapped, we call that a black hole and feature it in sci-fi movies.

The fabric of space and time is not a blank canvas on which the universe is painted. What is it, at a fundamental level, that makes up space in the universe? We don't know yet. What we do know is that space is more like a fluid that we move through without noticing but that is dynamically responsive to the presence of matter and energy. The more mass and energy there is, the more spacetime curves. The curvature of spacetime is what gives rise to gravity. And the more spacetime curves, the more it affects the motion of matter and energy. Each event, every fleeting moment, is akin to a ripple or current in a vast ocean. Spacetime has been stretching and warping since the beginning, a continuous series of explosions, expansions, and collapses (big bangs or big crunches, depending on if you're inside the action or on the outside, looking in). Matter and energy determine how warped the spacetime continuum is. And spacetime determines how the matter and energy will move. And the movement also warps spacetime, which further accelerates the matter/energy. And on and on.

When the big monsters of deep space (black holes) crash into each other, everything from galaxies to photons feel the waves. Maybe it's possible, or maybe not (but maybe yes!) that some of those big waves are so slow and so big that the night sky as seen and recorded by our ancestors was warped by these waves, and maybe they're still warping our vision today.

ENERGY FIELDS: EVERY POINT TELLS A STORY

What is a field? In a scientific sense, a field is something that has a particular value at every point in space. In our everyday lives, we're mostly familiar with electric (think of static electricity) and magnetic (as in a refrigerator magnet) fields. An electric field around a charged object assigns a value to every point in space, denoting the electric force that another charged object would experience if placed at that location. Similarly, a gravitational field describes the gravitational force that a mass would feel at every point in space. These fields allow us to understand and predict how objects will interact even when they are not in direct contact with each other.

Our modern understanding of physics is that there

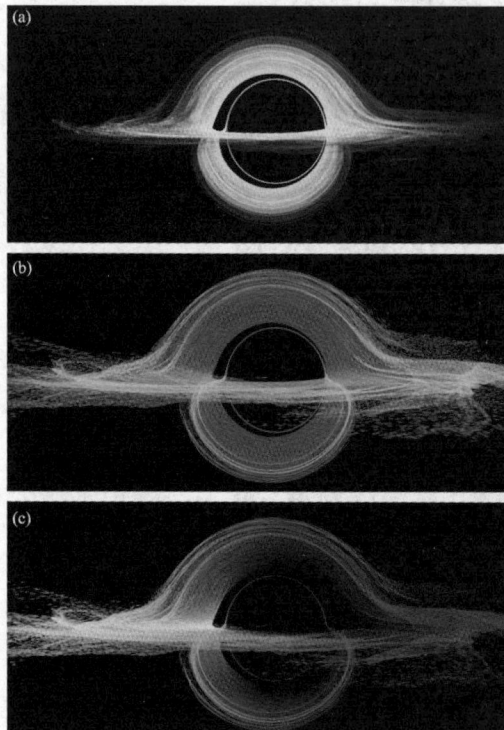

A moderately realistic accretion disk gravitationally lensed by a black hole (a) with its colors Doppler- and gravitationally shifted, (b) with the brightness adjusted, and (c) showing what the disk would truly look like to an observer near the black hole.

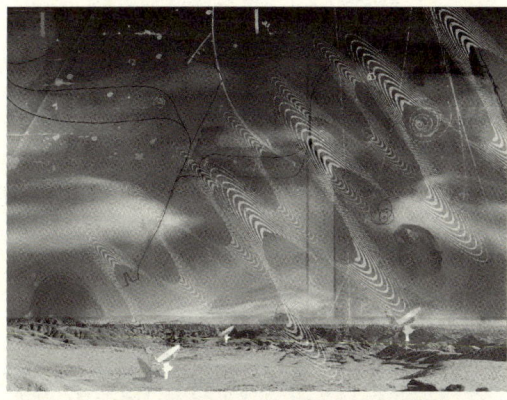

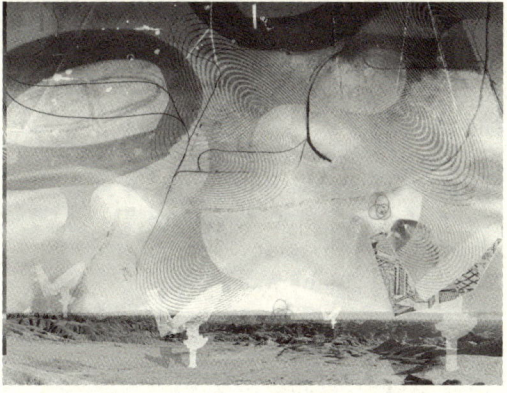

Eddie Edwards, *Galactic Ecosystems*, 2017. In June 2017, the National Radio Astronomy Observatory invited Edwards and three other artists to depict the workings of the next-generation Very Large Array, a radio interferometer planned for New Mexico.

are just four fundamental forces of nature: gravity, electromagnetism, the weak nuclear force, and the strong nuclear force. Already in that wording we've covered a major revolution of nineteenth-century physics: the idea that electric and magnetic fields are just part of the same effect. The nuclear forces operate only at the tiniest distances, mainly between subatomic particles.

A creature of subatomic size would sort of ignore electromagnetism and gravity, because the nuclear forces would be overwhelming. Similarly, for mosquitoes, electric fields are their whole world—gravity is a little drag, but for the most part everything feels sticky (i.e., static electricity is a major concern for the tiniest earthlings). For creatures the size of planets or stars, the field that matters is the one that connects us all: gravity. Gravity is the weakest force, but it is the one that rules the motions of the trillions of galaxies that make up the universe.

Even the existence of subatomic particles is due to fields: electrons, protons, neutrons, and more. They can all be thought of as distinct vibrations of some underlying field. The big bang was loud and nearly instantaneous, like the impulse of a drumstick. And just as whacking a string with a stick will excite all the notes and harmonics of that string, so did the big bang produce all the possible particles. Today, what we are left with are the notes that have not yet damped out.

For thousands of years, visible light was the only way to study the universe. Astronomers used telescopes to collect light from stars and galaxies, and they used this light to learn about their compositions, distances, and motions. All light is an oscillating electromagnetic field—all along the ray of light, the electric and magnetic field wiggles at a few hundred trillion times a second. When this radiation hits our retinas, it triggers biological circuits that our brain interprets as the real world. This light we see is not the only form of electromagnetic radiation. There are also radio waves, microwaves, infrared light, ultraviolet light, x-rays, and gamma rays. During the twentieth century, astronomers began to use these other forms of electromagnetic radiation to study the universe. They found that each form of radiation can reveal different information about objects in space—like the difference between black-and-white and color photographs, but a million times more profound. Radio waves can be used to study the hot gas rings around black holes, microwaves can be used to study the afterglow of the big bang, infrared light can be used to record creatures of the night and study the most distant stars, x-rays can be used to study the centers of teeth, bone, and galaxies, and gamma rays can be used to study the most energetic events in the universe, such as supernova explosions and neutron star collisions.

All the information we have ever received from nature is communicated by electromagnetic fields: light, sound, smell, and all the inputs of all of our scientific sensors. Light that we see with our eyes comes from oscillating electric and magnetic fields that propagate through space (we call this electromagnetic radiation), while gravitational waves originate from the warping of spacetime. What warps spacetime? Everything. All of us do it all the time. Each time we roll over in bed or do a little dance, we are sending gravitational ripples out into the universe. The ripples that our bodies make travel on across the universe forever.

These gravitational waves are like sounds that need no material medium to propagate. The frequency of gravitational waves, akin to the pitch of sound, largely depends on the mass of the objects causing them. The larger the mass involved and the slower it moves, the lower the frequency of the wave produced, and vice versa.

So, how do these waves affect light?

When gravitational waves pass through an area, they lead to a stretching and compressing of spacetime. Think of it like a subtle swell moving across the surface of a pond. Any light passing through those regions would be warped, somewhat like how light seems to bend going through water.

Gravitational-wave detectors, like the Laser Interferometer Gravitational-Wave Observatory (LIGO) in the U.S. and the Virgo interferometer in Italy, exploit this property using laser beams. These beams are fired into vacuum tubes, where they bounce between mirrors and, under normal conditions, come together and cancel out. This cancellation happens because the beams are "out of phase": one of the beams travels slightly farther than the other. When they bounce off mirrors and come back together, one of the waves is waving up while the other waves down, such that no light is visible. However, as a gravitational wave passes through the detector, the space inside of it expands and contracts, causing a tiny shift in the spacetime, turning the delicate balance of darkness into a brief shimmer.

Using all the tricks in the quantum measurement toolbox, LIGO detects gravitational waves with frequencies in the human audio range from mergers of black holes and neutron stars. On the other hand, measurements by the North American Nanohertz Observatory for Gravitational Waves (NANOGrav) are sensitive to ebbs and flows on the timescales of years, detecting the slow waves produced by supermassive black holes at the center of galaxies. NANOGrav is an astronomical project that uses the steady pulses from spinning stars (called pulsars) to detect gravitational waves. Pulsar timing arrays are made up

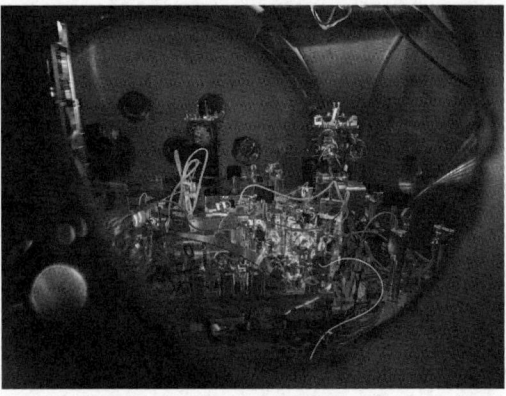

Part of the quantum engineering instrumentation used in the Laser Interferometer Gravitational Wave Observatory (LIGO). The green laser light is passed through a crystal which can reduce the infrared noise of empty space by fifty percent. Photograph by Georgia Mansell/LIGO Laboratory/Caltech.

of a network of radio telescopes that are used to precisely time the arrival of the pulses. When a gravitational wave passes through the galaxy, it causes the pulsars to appear to be pulsing slightly faster or slower than they are. By carefully timing the pulses and looking for subtle variations in the rhythm, NANOGrav can detect the presence of gravitational waves. These waves are stretching the whole Milky Way galaxy, and the pulsars are all acting as precise clocks. When the stretching ensues, an observer on the earth sees that all the "clocks" are displaying a different time. Both LIGO and NANOGrav have opened a new window on the universe, allowing us to study events that are impossible to see with other telescopes. Or so we thought.

THE HIDDEN SCIENCE IN ARCHAIC ARTISTRY

The universe's age, along with our planet's many geological epochs, demonstrates the vastness of time. These profound shifts in the universe's makeup, size, and shape happen so slowly that the adjective "glacially" means "speedy." Consider this: a 100-year-old human's lifetime is approximately three billion seconds. While that might seem like a long time, when juxtaposed with the multi-billion-year age of the universe, it's the briefest flash of lightning. While it's true that our individual life spans are short, our collective history as observers of the cosmos extends much further. It would be hard to point to a time in our evolution when we were completely disconnected from the night sky.[2]

The vast expanse of the night sky was a muse to humanity long before we penned our first word or erected our first monument. Though the clearest evidence of ancient astronomy is found in written documents, oral narratives, visual symbols, and the strategic alignment of specific architectural structures, a growing body of evidence suggests that it was not unique to the Scientific Revolution or to the rise of ancient civilizations.[3] Astronomy was an integral part of human culture long before the advent of written records. During the Stone Age, the hunter-gatherers who roamed the earth were aware of the solar movements, lunar cycles, stellar positions, and corresponding changes in native flora and fauna. For them, the sky above and the life below were woven together, informing and shaping their lives. In other words, there was a belief in a concordance that allowed them to look at the sky and predict what would happen on the surface of the earth.

Before the emergence of written records, practices that aided survival, such as seasonal forecasting, food procurement, and celestial navigation, were passed down through generations using art and oral traditions. Stories, songs, images, and movement not only fed the imagina-

2. People growing up in highly light-polluted cities are cut off from the direct experience of a night sky, although they can get it vicariously through recorded images. This may be the first time in our species' history where some of the population's formative years happen in the complete absence of astronomical majesty.

3. Brian Hayden and Suzanne Villeneuve, "Astronomy in the Upper Paleolithic?," *Cambridge Archeological Journal* 21, no. 3 (October 2011): 331–55.

tions and souls of our ancestors, but also encoded celestial observations and wisdom necessary for the survival and wellbeing of human societies.

Ancient humans arrived on the continent of Australia roughly 65,000 years ago, inhabited Eurasia and Oceania at least 50,000 years ago, and 15,000 years ago populated the Americas with memories and stories of fire and the stars.[4] They explored the seas and the lands with the twin flames of creativity and curiosity. The night sky was a grand canvas on which stories of gods, heroes, and monsters were painted. Constellations were not just clusters of stars; they were mythological narratives, each with its own tale. Because of their dependability, the moving lights that traversed the heavens often came to be regarded as ancestor gods. Little wonder that celestial objects appear frequently in oral and written mythologies that tell the story of creation and the path of humanity and the lineage of the storyteller from the sky.[5]

The Dunhuang star atlas is one of the most spectacular documents in the history of astronomy. It is a finely detailed representation of the Chinese sky, including numerous stars and asterisms, depicted in a succession of maps. Apart from its aesthetic appeal, the document, found on the Silk Road, is remarkable because it is the oldest star atlas known today from any civilization.[6] Scholars consider it a syncretic document that amalgamated information from multiple sources into an early scientific perspective that also served for divination.

The Australian Aboriginal cultures gazed upward and discerned their own celestial stories. The Emu in the Sky is a famous dark constellation, made not of stars but of the dark spaces in the Milky Way. Its appearance in the sky signaled the time for emu egg collection. Beyond myth and storytelling, these cultures used the night sky as an essential tool to develop complex "songlines," or dreaming tracks, which were navigational paths across the land and sky intertwined with stories and ceremonies. The Bhimbetka rock shelters in India contain around 500 paintings strikingly like those in Australia's Kakadu National Park as well as to the San people's cave art in the Kalahari Desert and the Upper Paleolithic murals of France's Lascaux cave.

In New Zealand, the Māori relied heavily on the stars for navigation and timing. Matariki, the Māori name for the Pleiades star cluster, heralds the start of the New Year in the Māori calendar. Its rise before dawn typically marks a time of reflection, celebration, and preparation for the year ahead. The Polynesians, with their navigational prowess, roamed the Pacific using the stars as their compass. Wayfinders would set sail using constellations, the rising and setting points of stars, and celestial bodies

4. Stephen Oppenheimer, "Out-of-Africa, the Peopling of Continents and Islands: Tracing Uniparental Gene Trees across the Map," *Philosophical Transactions of the Royal Society B: Biological Sciences* 367, no. 1590 (March 2012): 770–84.

5. Anthony F. Aveni, "Archaeoastronomy in the Ancient Americas," *Journal of Archeological Research* 11, no. 2 (June 2003): 149–91.

6. Jean-Marc Bonnet-Bidaud, Françoise Praderie, and Susan Whitfield, "The Dunhuang Chinese Sky: A Comprehensive Study of the Oldest Known Star Atlas," *Journal of Astronomical History and Heritage* 12, no. 1 (2009): 39–59.

7. Manuel León-Portilla, "A Reflection of the Ancient Mesoamerican Ethos," in Aveni, ed., *World Archaeoastronomy* (Cambridge, England: Cambridge University Press: 1989), 225.

8. A. L. Peratt and W. F. Yao, "Evidence for an Intense Solar Outburst in Prehistory," *Physica Scripta* 2008, no. T131 (December 2008): 13.

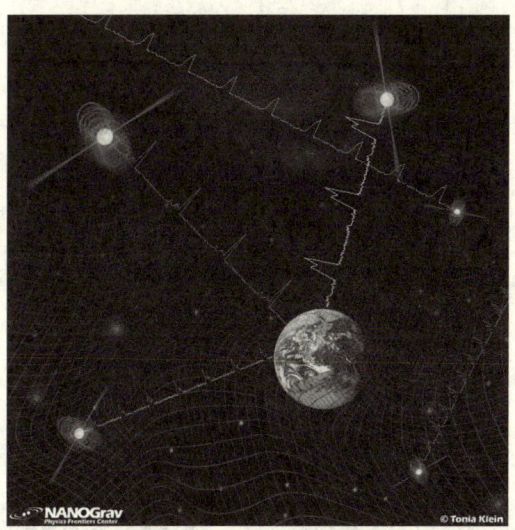

This artist's interpretation shows how radio waves from an array of pulsars are affected by a gravitational wave produced by a supermassive black-hole binary in a distant galaxy. Photograph by Tonia Klein for NANOGray.

like Venus and Jupiter to move between distant islands. Even the Chumash of present-day California looked to the heavens to guide them across vast oceanic and terrestrial expanses. These navigational techniques have existed in some form for a few thousand years, passed down through the generations via storytelling, apprenticeships, and traditions and ceremonies that survive in contemporary culture. No textbooks.

Agriculture, too, was guided by stars. From the flooding of the Nile in Egypt, cued by Sirius, to planting and harvesting in Maya according to the cycles of the Pleiades, and even to the timing of agricultural festivals in India and the Pacific Islands, the rhythm of agriculture often danced to the celestial drumbeat. Torreón of Machu Picchu, Sun Dagger of Chaco Canyon, and the observatories of ancient Polynesia are grand gestures of stone and intent, aligning with solstices, equinoxes, and specific star patterns, revealing an intricate understanding of cosmic cycles.

The Aztecs said of Teotihuacan that it was the place where time began. Anthropologist Manuel León-Portilla has written that one's very existence in Mesoamerica depended upon observing the sky: "Without skywatchers, the ethos of this people, its distinguishing spirit, its own genius would not have developed."[7] The heavens were not just a source of wonder, but a compass, calendar, and chronicle all in one.

In ancient North America, petroglyphs, geoglyphs (human made formations inscribed on the ground), and pictographs carved by Indigenous peoples portray the sun, the moon, and star patterns. These ancient etchings provide insight into cosmological views and attempts to understand celestial phenomena like solar eclipses, aurorae, and meteor showers. A survey of petroglyph sites encompassing some four million pieces of rock art in 139 countries found evidence for an intense solar outburst in prehistory.[8]

As time wore on, our media evolved. Island cultures have long been captivated by the night sky. From Japanese artists capturing constellations in woodblock prints to Polynesian chants referencing celestial bodies, these artistic expressions showcase the enduring fascination with the heavens.

In light of these many artistic representations and cultural inscriptions that convey information about the natural world, one is compelled to question: What does it truly mean to make a scientific observation? Traditional science often emphasizes detached objectivity and precision. But when early civilizations painted, sang, or carved their celestial observations, they were not merely noting down data points. They were intertwining observation with

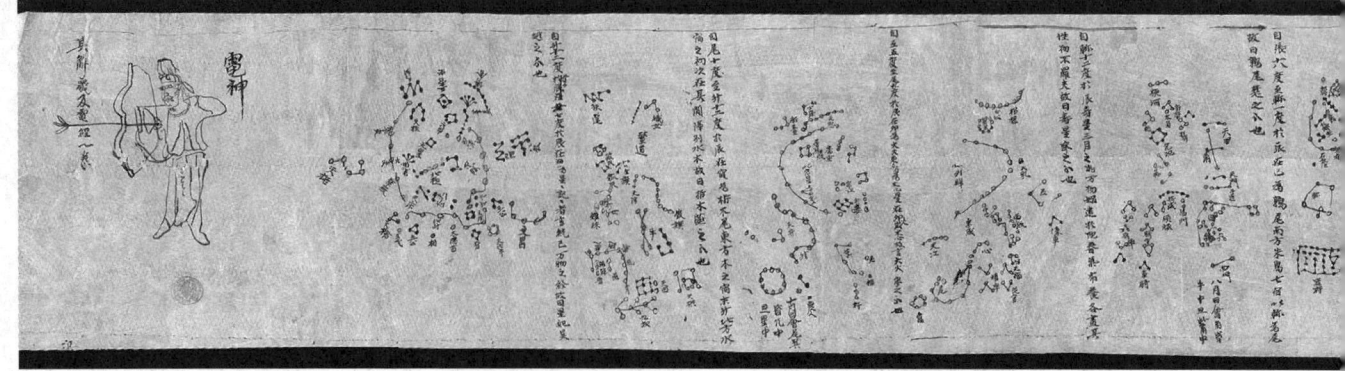

meaning, ensuring that the knowledge would not only be stored but felt, remembered, and passed down. By embedding these cosmic patterns and phenomena within our stories, art, and myths, they became more than raw data; they became part of our shared human identity. It's a testament to the idea that scientific observation isn't merely about quantifiable data—it's about embedding knowledge into the cultural fabric, ensuring its preservation and resonance for generations to come. It's about making sure the information survives.

IT FROM BIT

Across both space and time, shared experiences have knit humanity together, manifesting in the mosaic of our civilizations. Information was encapsulated in rhythms, rhymes, stories, and songs long before the digital age. These mediums served not merely as forms of entertainment or artistic expression, but as crucial tools for preserving knowledge.

  The twentieth century heralded the era of information theory and computing, addressing the necessity of storing vast amounts of information without degradation. This led to the development of algorithms that could compare multiple strands of data to identify and correct random errors, a precursor to the sophisticated error correction methodologies we see today. Contemporary science, particularly in the realm of quantum computing, is deeply concerned with the idea of error correction: how to keep all the little bits in your conventional or quantum computer from getting flipped by electrical interference or cosmic rays. This concept echoes ancient methods of error correction, where a song's rhythm or a story's trope helped maintain the information's integrity, drawing a poetic parallel between the past and the present.

  John Archibald Wheeler's "it from bit" theory is a concept in the philosophy of physics exploring the fundamental nature of reality, particularly in the context of quantum mechanics and information theory. The essence of Wheeler's idea is captured in the phrase "it from bit," suggesting that the fundamental nature of the physical

The Dunhuang star atlas, from seventh-century China, is the oldest known graphical star map. Its stellar positions were so accurate that 400 years passed before a revised chart appeared.

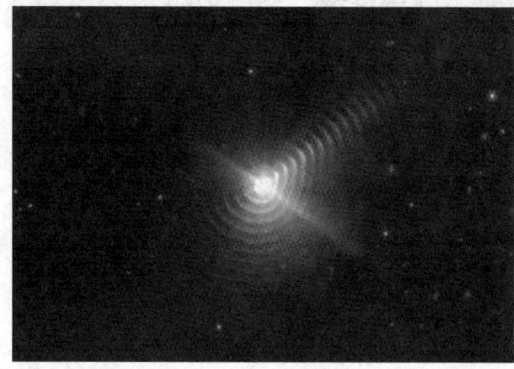

Two Wolf–Rayet stars orbiting around each other produce shells of dust every eight years that look like rings, as seen in this image from NASA's James Webb Space Telescope. Photograph courtesy NASA, ESA, CSA, STScI, JPL-Caltech.

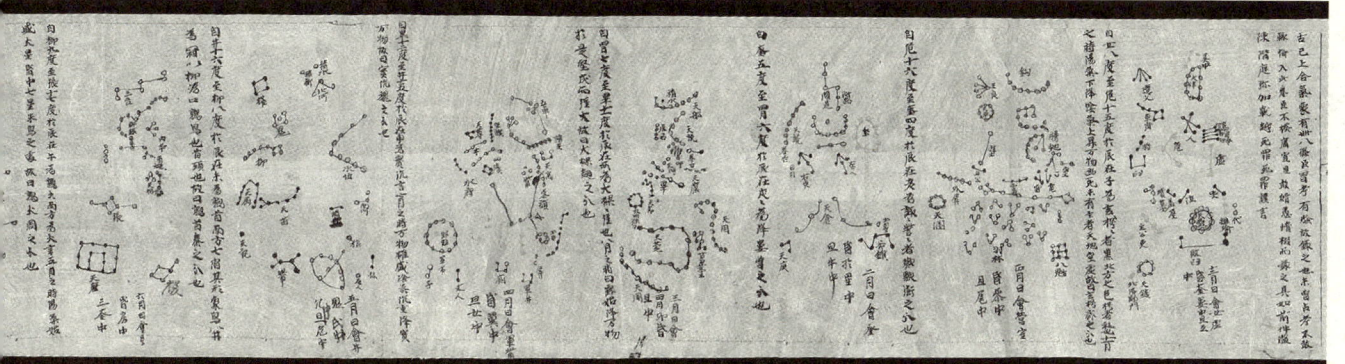

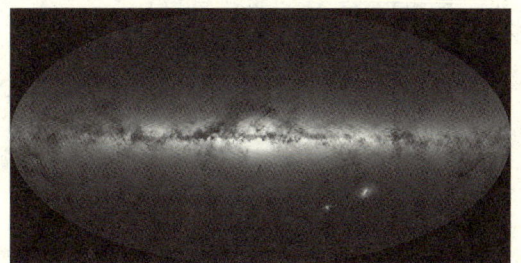

An image of the night sky as observed by the European Space Agency's Gaia spacecraft. The bright stripe through the middle is our own Milky Way galaxy. Photograph © ESA/Gaia/DPAC.

universe (the "it") arises from the fundamental aspects of information (the "bit").

Our ancestors, acting as proto–information scientists, encoded wisdom in formats resilient against the ravages of time and the fallibility of human memory. As we navigate through our information-dense digital age, the ancient ingenuity in safeguarding and perpetuating knowledge through stories, songs, and art continues to resonate. The enduring quest for information preservation, embodied in every beat, verse, and brushstroke from our past, reflects our intrinsic drive to understand, convey, and interact with the foundational language of the universe.

THE INFORMATION

What is the real meaning of those [something] vs. time plots that scientists show?

As an example, we might ask about figures showing the global average temperature on the vertical axis of a plot and time on the horizontal axis. What that visual does is describe the evolution of something over some period not as a changing thing but as a fixed squiggle on a piece of paper. The dimension of time is represented as the dimension of space: the horizontal axis.

This is just an imitation of something we have seen in nature: the thickness and character of tree rings are the dimension of time made visible as space, marching out from the origin of the tree at the center. Every tree ring tells its own story, with its width indicating the water availability for that year. By gathering and analyzing these distinct patterns from numerous trees across various regions, scientists can gain insights into historical climates, ecosystems, and even bygone civilizations. Similarly, we see the passage of geologic time in the relative coloring of rocks; a recording of what kinds of "stuff" was in the air at that time. We use the growing tree or the developing rocks as the recording medium upon which the pen describes what nature was doing.

Space, however, is big. The human eye can see the light from distant stars—so distant that the light left those stars a million years ago, while our ancestors were just forming the beginnings of our civilization. Just as a pebble makes a circular wave in a still pond, galactic-scale events in the distant past might still be rippling throughout the night sky. It might seem as if we were gazing in awe at an image of the sky reflected from a pond rather than at the sky itself. Today, as astronomers record ever more precise images of the night sky, they can look for small motions over the span of years or even decades. By comparing observations made thousands of years ago with tonight's sky, we can construct a better representation of our evolving universe and try to understand why nature is the way that it is.

The most famous star cluster in the Northern and Southern Hemisphere is the Pleiades (also known as the Seven Sisters, or as Subaru in Japan). As highly visible heliacal stars, the Pleiades were observed and used by many cultures around the world, including the Māori, Aboriginal Australians, Chinese, Mayan, Aztecs, Vedic Hindus, Indigenous peoples of North America, and ancient Greeks. Given the similarities in myths about the Pleiades across different cultures, it could be that the Seven Sisters' story predates the departure of the ancestors of Aboriginal Australians and Europeans from Africa more than 100,000 years ago. It may even be the world's oldest story, also making it the longest continuous record of astronomical data.[9]

It's fascinating to consider that the human eye, long before the invention of telescopes, was such a proficient tool. Capable of discerning thousands of individual stars and even detecting faint celestial objects, our ancestors had a nightly spectacle of cosmic wonders. They relied on their eyes, and their sharp observational skills, to understand and map the movements and stories of the heavens. Over millennia, these stars haven't remained static. All stars are moving somewhere on their own path at their own speed. Their individual undulations are written on an underlying gravitational scaffolding erected by some sort of invisible (dark) matter or field that modern astronomy has yet to understand. But for over 50,000 years, humans have been observing the night sky with their keen farsighted vision and clear skies. If we could put all these observations together, we would have a wondrous movie. Sped up by a billion times, we could relax on a couch with some popcorn and watch the history of the universe unfold.

This is essentially what we have in the physical and cultural records left on earth by our ancestors. Not at the Hubble Space Telescope level of accuracy—but information, nonetheless. By juxtaposing ancient portrayals of the night sky with our current understanding, we can discern how stars and constellations have moved or evolved.

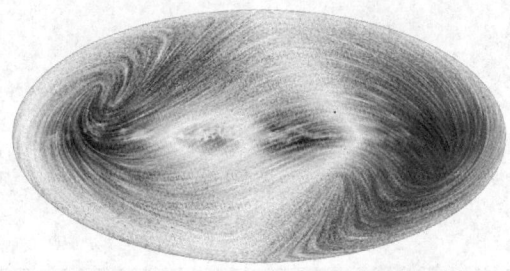

This image from the European Space Agency's Gaia spacecraft shows the speed at which more than thirty million Milky Way stars are moving towards (blue) or away from (red) Earth. Photograph © ESA/Gaia/DPAC.

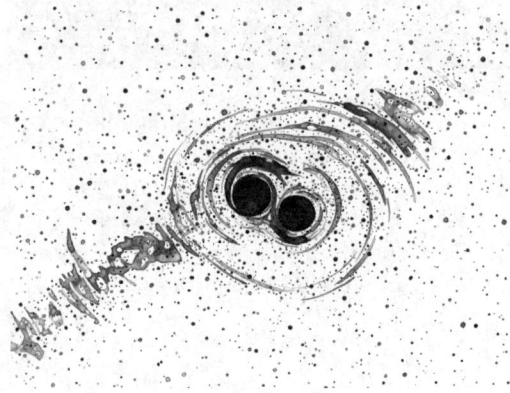

Kip Throne and Lia Halloran, *Colliding Black Holes*, 2023. Ink on duralar, 17 x 22 in. Photograph by Paul Salveson, from *The Warped Side of Our Universe* (New York: Liveright, 2023).

9. Ray P. Norris and Barnaby R. M. Norris, "Why Are There Seven Sisters?," in *Advancing Cultural Astronomy: Studies in Honour of Clive Ruggles* (New York: Springer, 2021), 223–35.

10. "Astrometry," in *History of Astronomy: An Encyclopedia*, ed. *John Lankford* (Abingdon, England: Routledge, 1997), 49.

This depends critically on the accuracy and precision of observations made by eye and the manner in which they are recorded, whether as carvings and drawings or as stories and songs. As described by John Lankford, before the invention of the telescope, over 500 years ago, people were able to measure the positions of stars with an accuracy of around twenty arcminutes (around one- quarter the size of the full moon).[10]

In recent years, astronomers have started to make very detailed pictures of the night sky with so much precision that we may be able to directly see the warping of spacetime in images of the galaxy. It may yet be a few decades before we achieve that kind of resolution, but once we do, we will be able to watch the constellations slowly breathe, like goldfish viewing their own bowl, looking up at the world through the waves. Evaluation of significant prehistoric folklore and ancient artifacts along with modern astrometric data could help measure what hasn't been detected yet—ripples in the fabric of space and time made by gravitational waves that emanate from the collisions of galaxies and supermassive black holes.

DECADES, MAKE THEM MATTER

The human eye, despite its remarkable capabilities, has its limitations. Our ancestors, lacking today's advanced technologies, blended observation and interpretation in their depictions, encapsulating not merely the stars, but the narratives and beliefs through which they understood nature. These ancient renderings reveal a record of the sky intertwined with the human passion of passionate inquiry. Today, as we gaze skyward with sophisticated instruments and methodologies, we build upon this rich historical foundation. The interplay of past and present observations not only expands our knowledge base but exemplifies the persistent arc of scientific inquiry throughout history.

Ancient civilizations, through their intricate artworks and narratives, have offered us invaluable data points that augment our contemporary astronomical observations. As we step into a technologically advanced era, artificial intelligence is playing an instrumental role in analyzing and interpreting enormous data libraries, furthering our astronomical understanding at unprecedented scales. By the end of the century, we may just be the consumers of secondhand information, rather than the ones making the discoveries. Understanding the understanding of our ancestors may be essential to smooth our own transition into ancestorhood. It is vital to preserve the best of humanity's wisdom and pass it on to whomever will need it in another 10,000 years. <

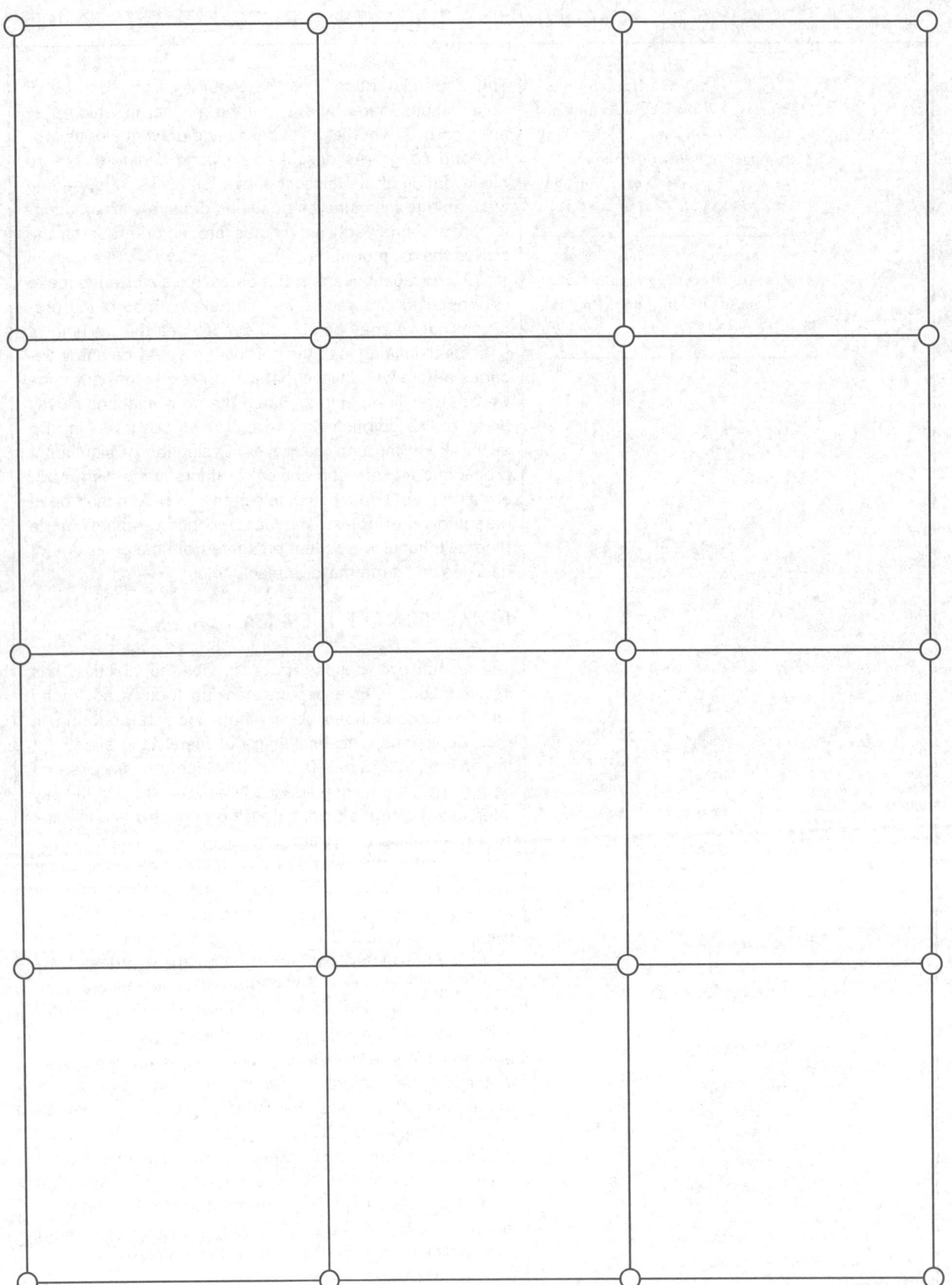

ALBA TRIANA

Since the early 2000s, Colombian artist Alba Triana has striven to understand the essence of things through vibration. She believes vibrational phenomena are fundamental to everything, and that everything possesses an inherent vibratory signature, from the most minuscule objects to the unfathomably vast universe. Trained as a composer, Triana expanded her practice to accommodate these explorations, incorporating sculpture, performance, and installation to produce immersive experiential environments. Her work invites the audience to encounter vibration and its effects in new ways.

Triana's *Music on a Bound String No. 2* (2015) is a kinetic sound and light sculpture that operates according to the principles of wave behavior and diffraction. A string, suspended between two points within a beam of light, is subjected to an infrasound audio signal. The signal is inaudible to the viewer, but it excites the string, causing it to oscillate so that it visually manifests a sound wave. As the string vibrates, it diffracts the colors that make up the beam of light, making various tones visible according to the dictates of its sonic agitation and the changes in the beam's spectral content. <

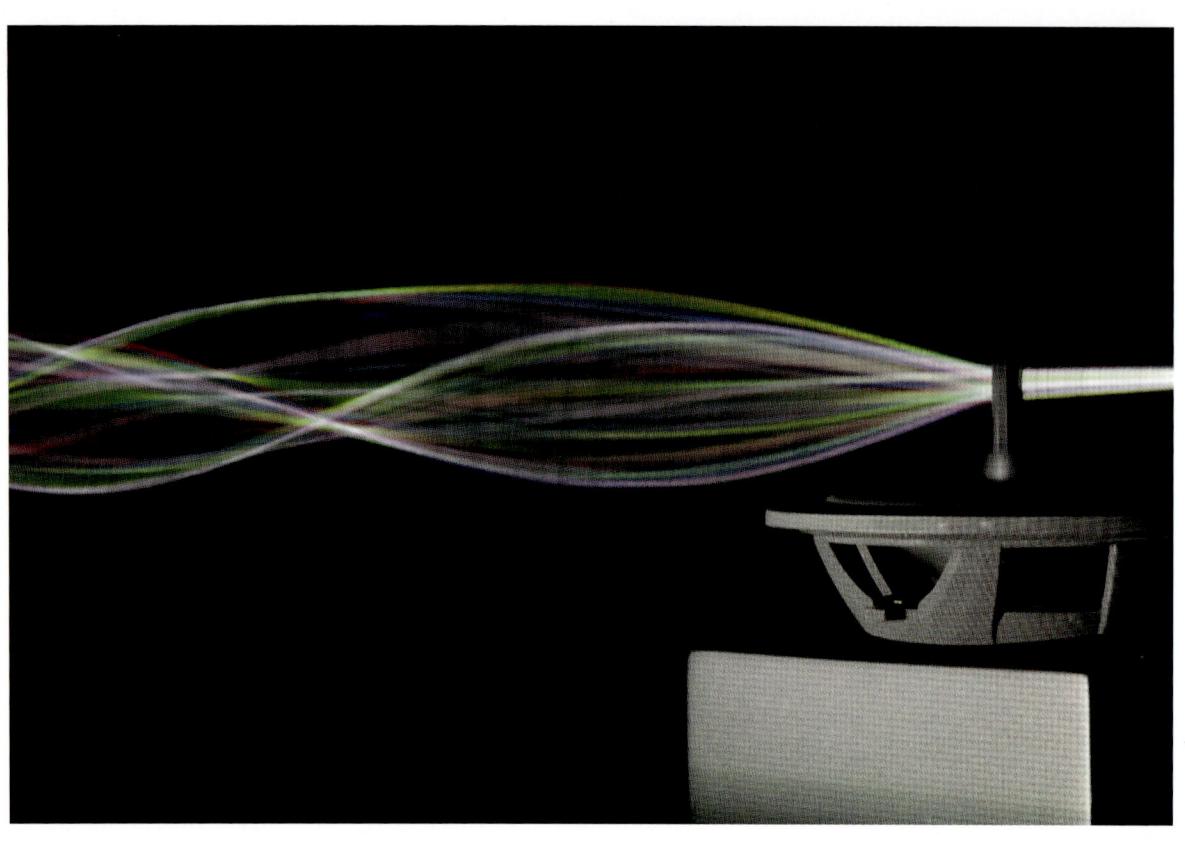

Alba Triana, *Music on a Bound String No. 2*, 2015.
Photographs by Ernesto Monsalve (above),
Silvia Ros (opposite, top), and
Bernardo Olmos (opposite, middle and bottom
and following page),
courtesy Alba Triana Studio.

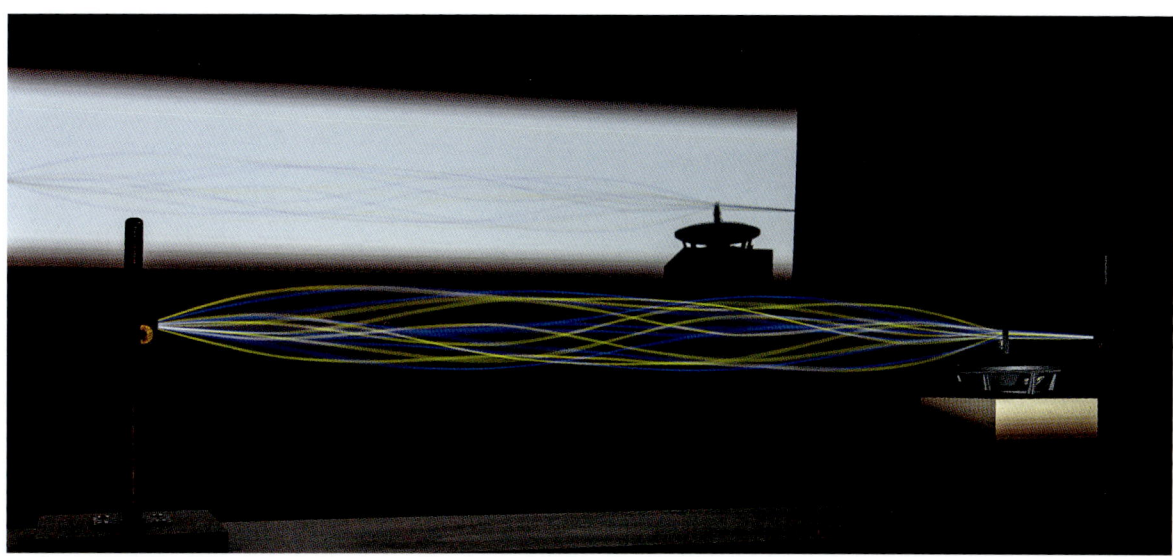

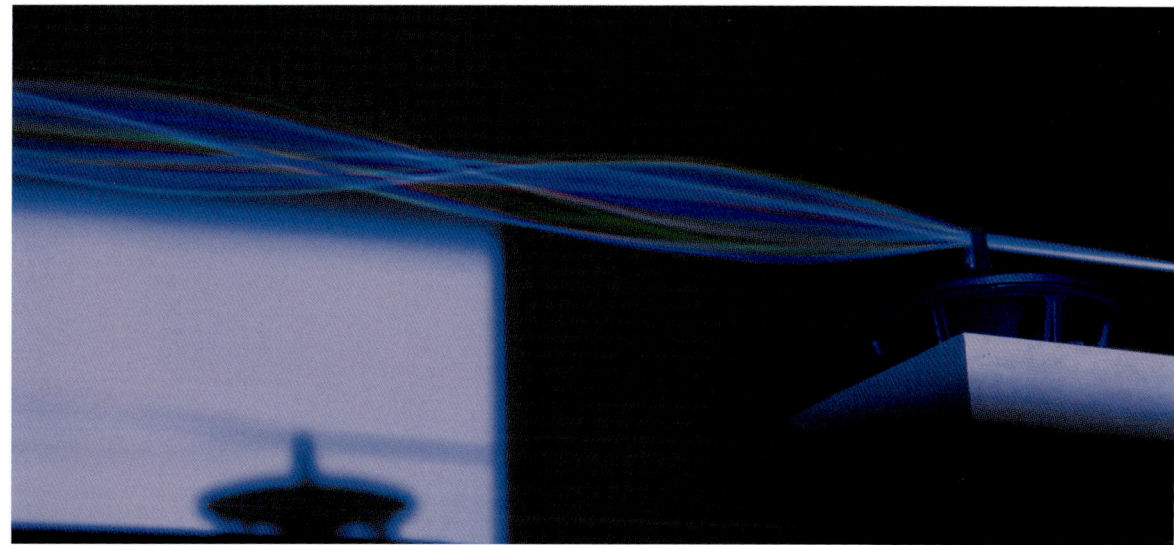

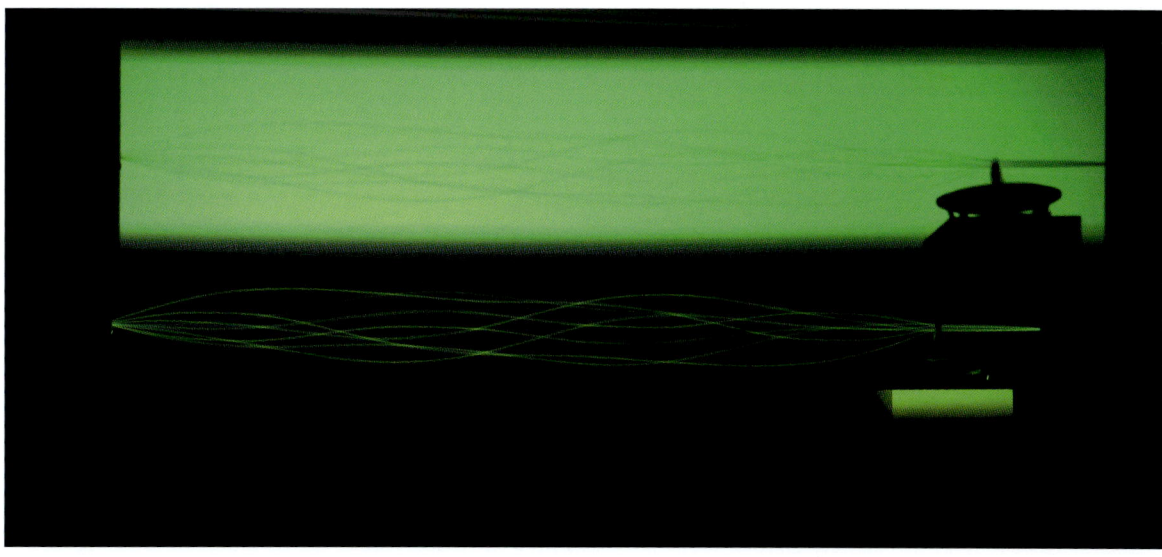

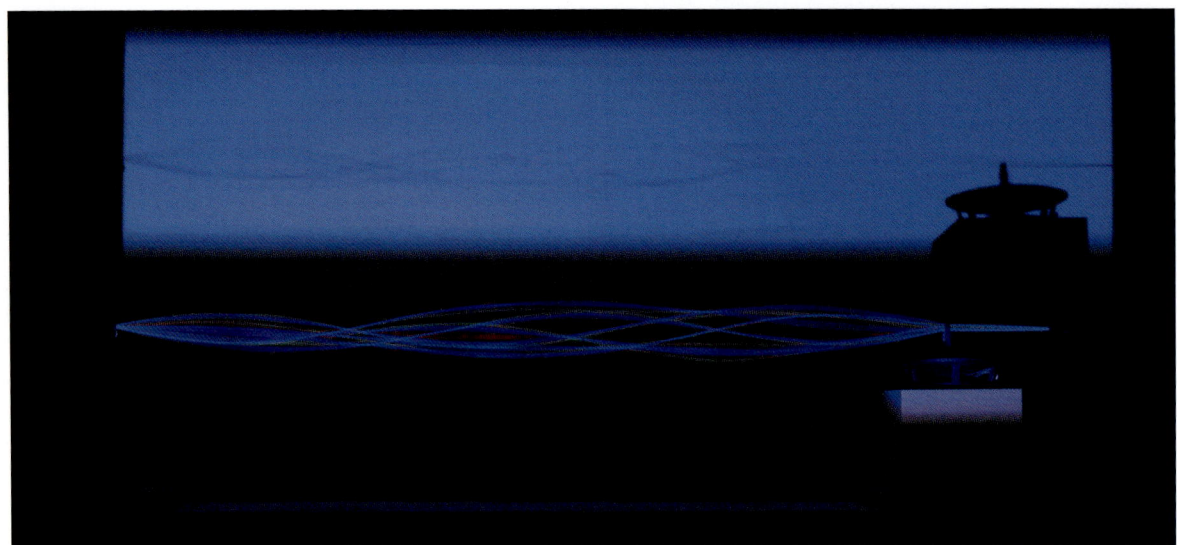

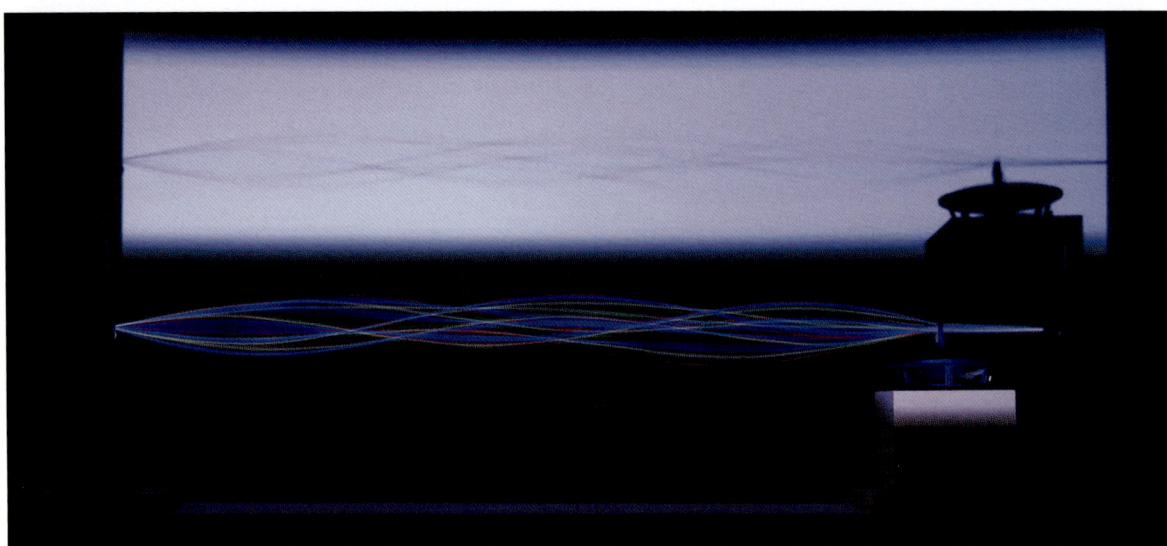

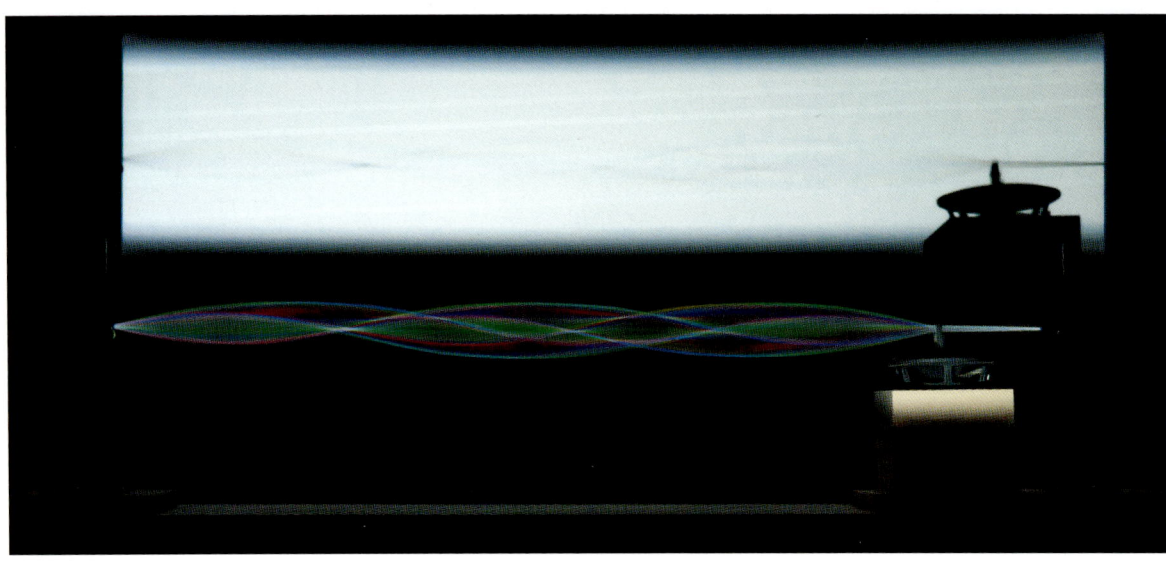

ROSS MANNING

Emerging in the early 2000s as an experimental musician, Brisbane-based Ross Manning has shifted his artistic focus from performance to installation. He has also moved from playing instruments with obvious interfaces to working with seemingly inert materials from which sounds and experiences must be coaxed. His installation work often embodies processes of transference and translation, harnessing invisible energies and rendering them into soundscapes, refracted light, or moiré patterns.

Manning's Ambient Paintings (2016 – ongoing) series channels ambient light to create color and patterns on the surface of blank canvases. These "paintings," which are variously sized, are composed of dichroic glass filters mounted to canvas at ninety-degree angles. As ambient light hits the glass filters, it splits into different wavelengths, which then hit the surface of the canvas, creating color and pattern. As the quality and intensity of the ambient light changes, the painting's composition changes; in this way, the work is a simultaneously reductive and expansive reading of the space, moment to moment. By rendering the electromagnetic spectrum visible, the Ambient Paintings prompt viewers to consider the spectral nature of light. <

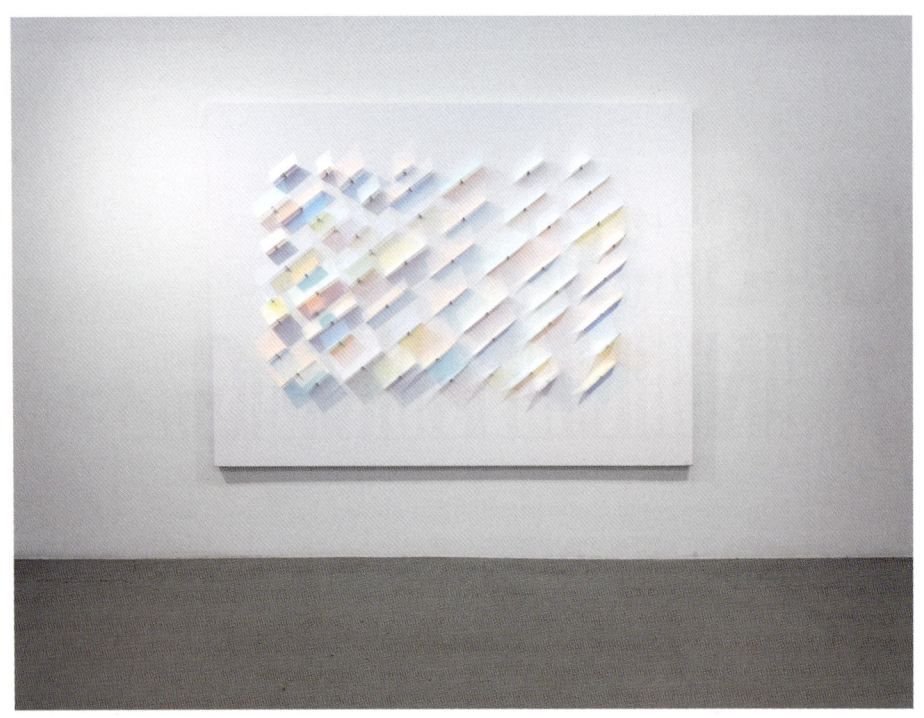

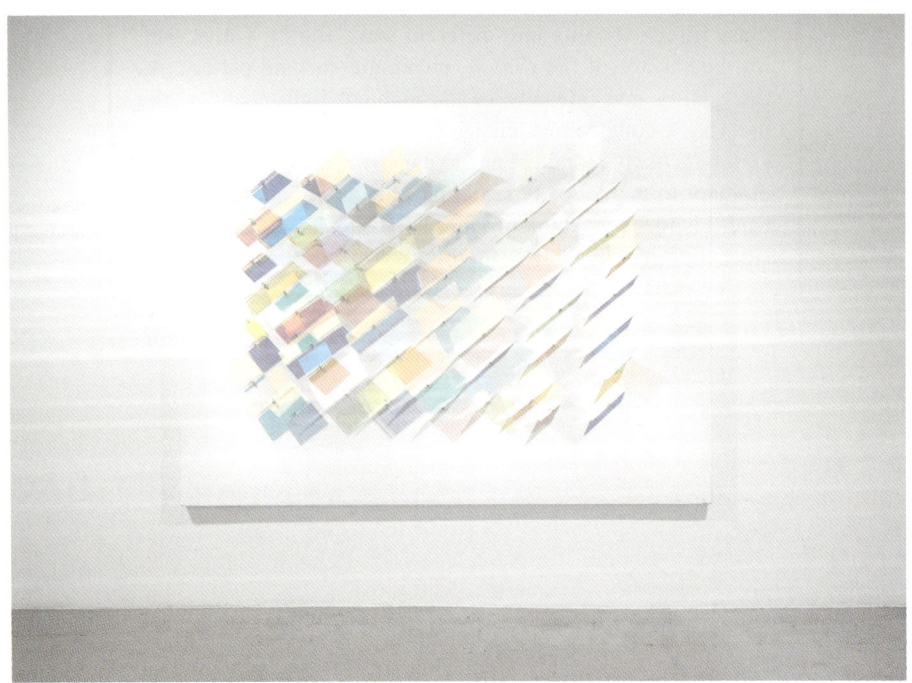

Views of Ross Manning's *Ambient Painting (Diagonal)*, 2021. Acrylic, stainless steel, and dichroic glass on canvas, 70 ⁷/₈ x 94 ¹/₂ in. Courtesy the artist and Milani Gallery. Photograph by Carl Warner.

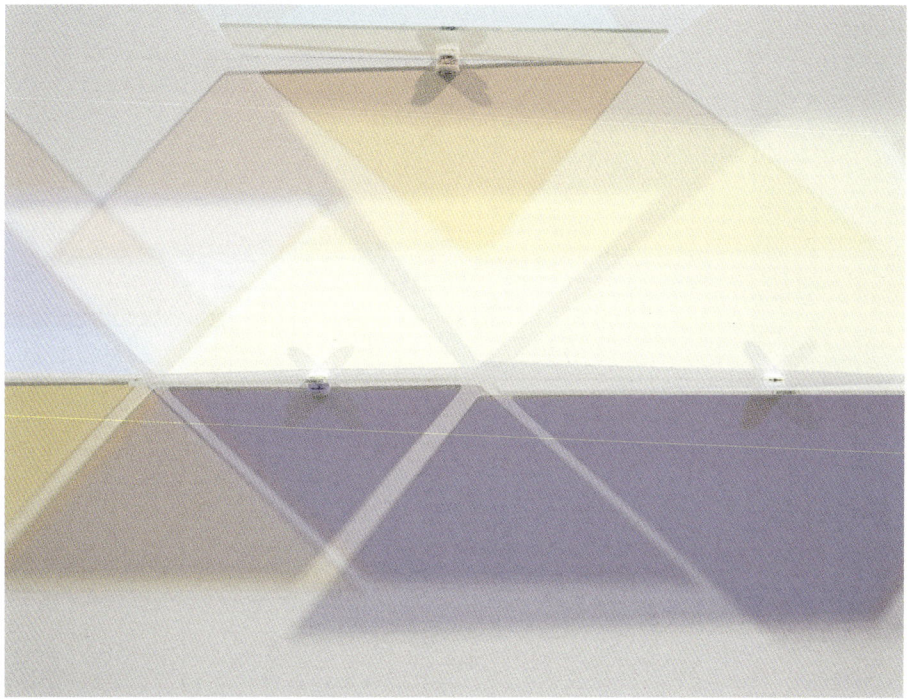

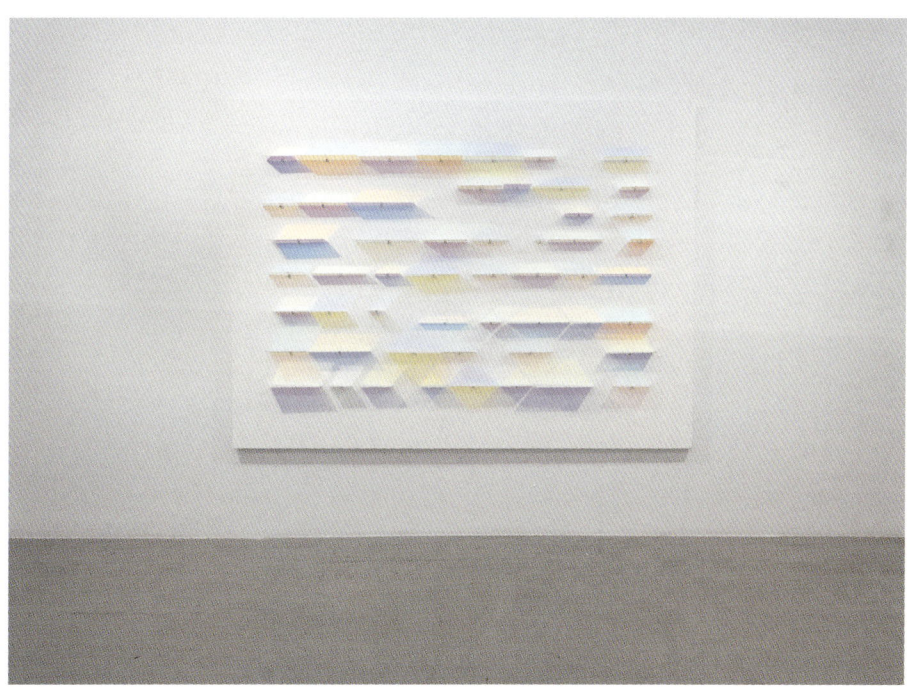

Views of Ross Manning's *Ambient Painting (Horizontal)*, 2021. Acrylic, stainless steel, and dichroic glass on canvas, 70 $^7/_8$ x 94 $^1/_2$ in. Courtesy the artist and Milani Gallery. Photograph by Carl Warner.

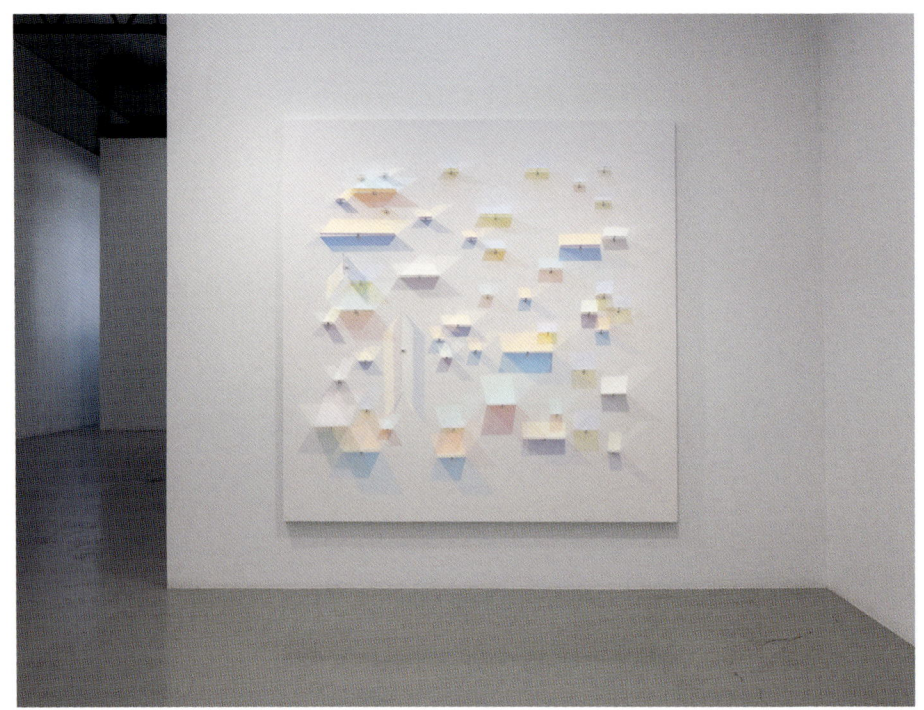

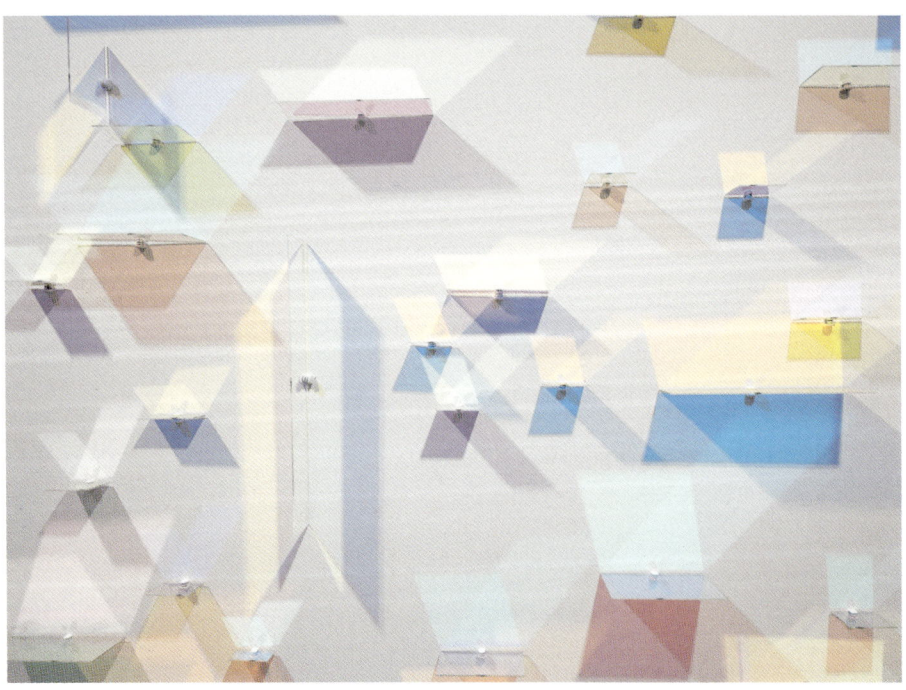

View and detail of Ross Manning's *Ambient Painting (Square)*, 2021. Acrylic, stainless steel, and dichroic glass on canvas, 78 ³/₄ x 78 ³/₄ in. Courtesy the artist and Milani Gallery. Photograph by Carl Warner.

LEN LYE

Born in 1901 in New Zealand, Len Lye lived in Australia and England before relocating to the United States in 1944. Fascinated with motion dynamics, he created pioneering films such as *A Colour Box* (1935) and *Free Radicals* (1958) by painting and scratching directly onto film strips to animate abstract patterns of color and light. This technique set a new precedent in avant-garde cinema. From around 1960 onward, his exploration of movement extended beyond filmmaking to include kinetic sculpture; works like *Bell Wand* (1965) and the posthumously erected *Wind Wand* (1996), made of flexible fiberglass, are designed to vibrate and bend as external forces act on them.

Lye's film *Particles in Space* (1980), presented in *Energy Fields: Vibrations of the Pacific*, explores the direct animation methods used in *Free Radicals*. The resulting dots and dashes flash and race across the screen with the kind of high-energy excitement that is a hallmark of the artist's work. The Len Lye Foundation notes that the artist "made several versions of the film in the 1960s but almost completely remade it in 1979 with the assistance of Steve Jones and Paul Barnes. Synchronized to drum music from the Bahamas and from Nigeria, the film begins and ends with sounds of Lye's steel kinetic sculptures (another area of art in which he was a pioneer)." Lye, who passed away in 1980 (the year *Particles in Space* was released), left a legacy affirming the generative potential of merging art and science, and his works continue to influence contemporary art and film. <

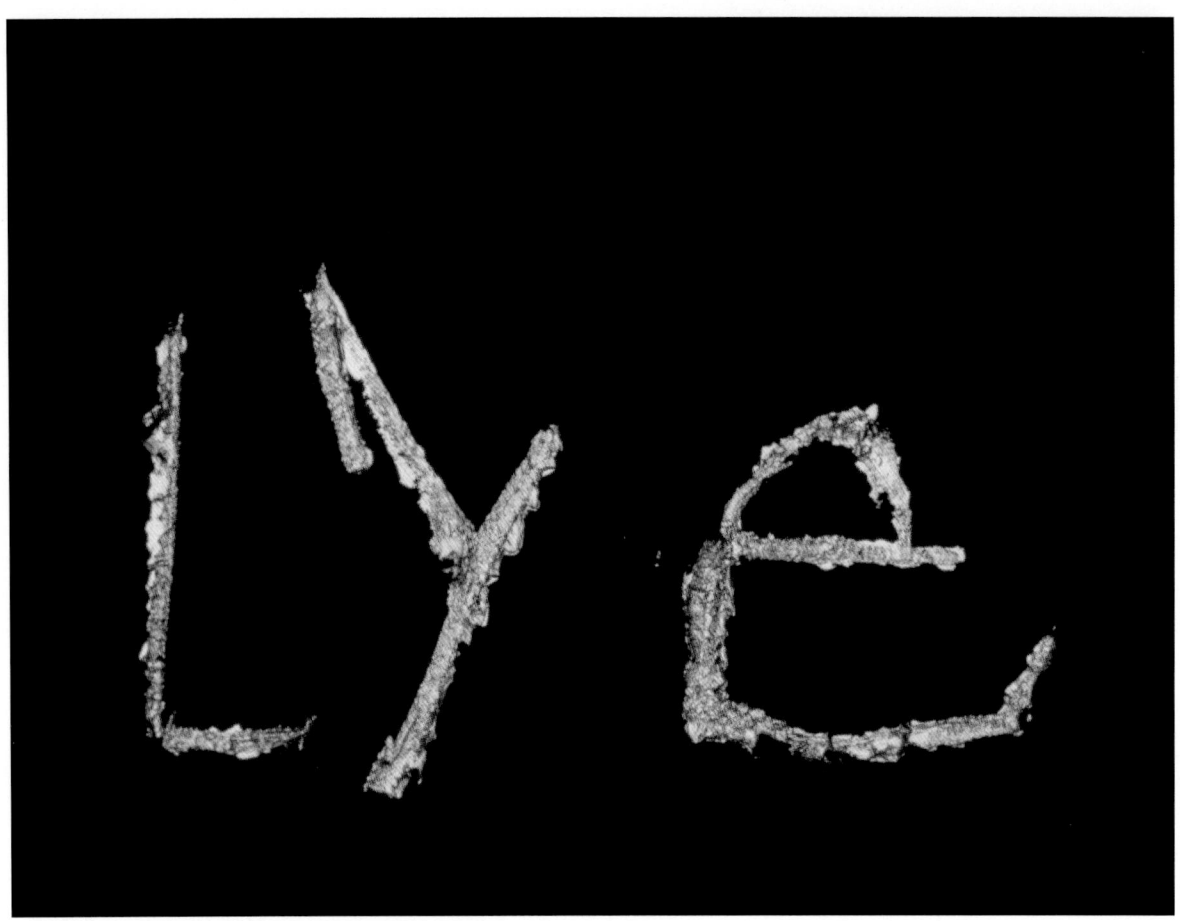

Len Lye, still from *Particles in Space*, 1980.
Photograph courtesy Ngā Taonga Sound
& Vision.

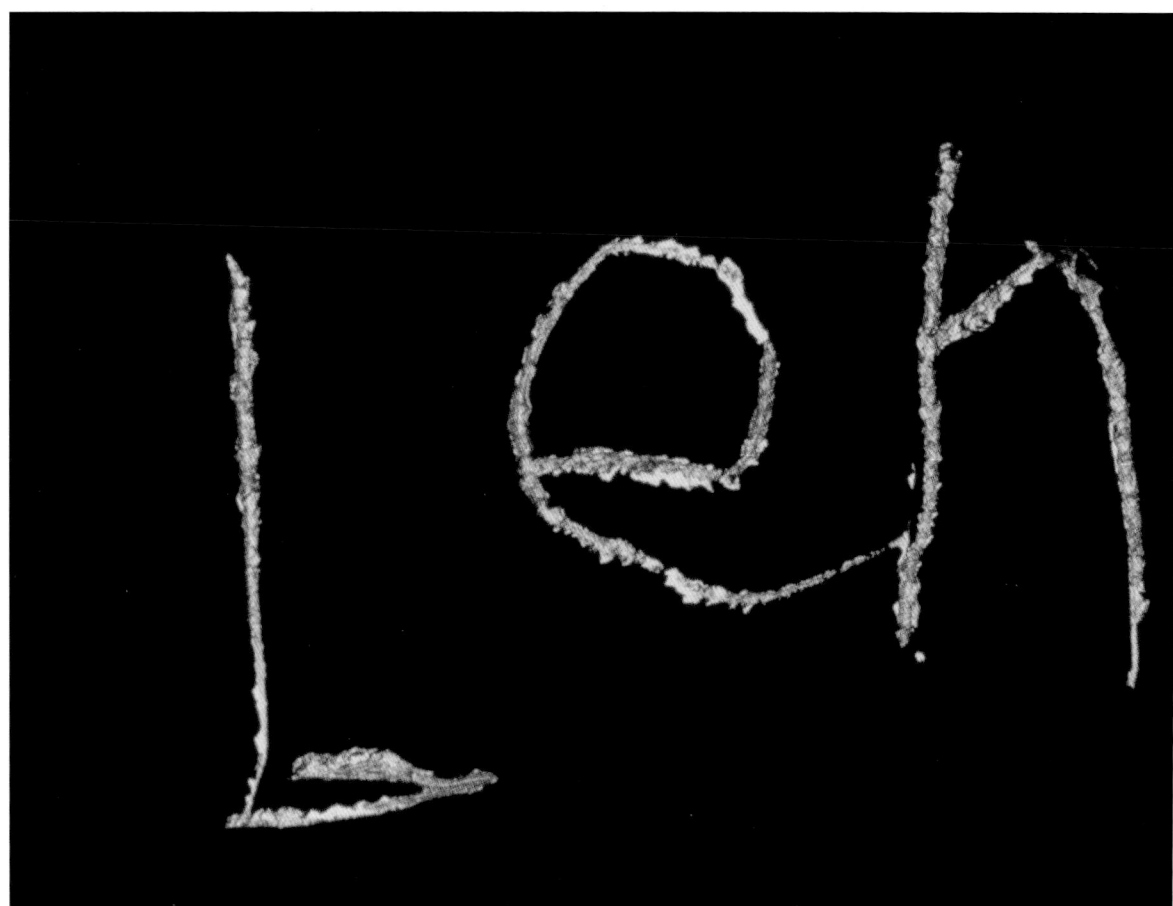

Len Lye, still from *Particles in Space*, 1980.
Photograph courtesy Ngā Taonga Sound
& Vision.

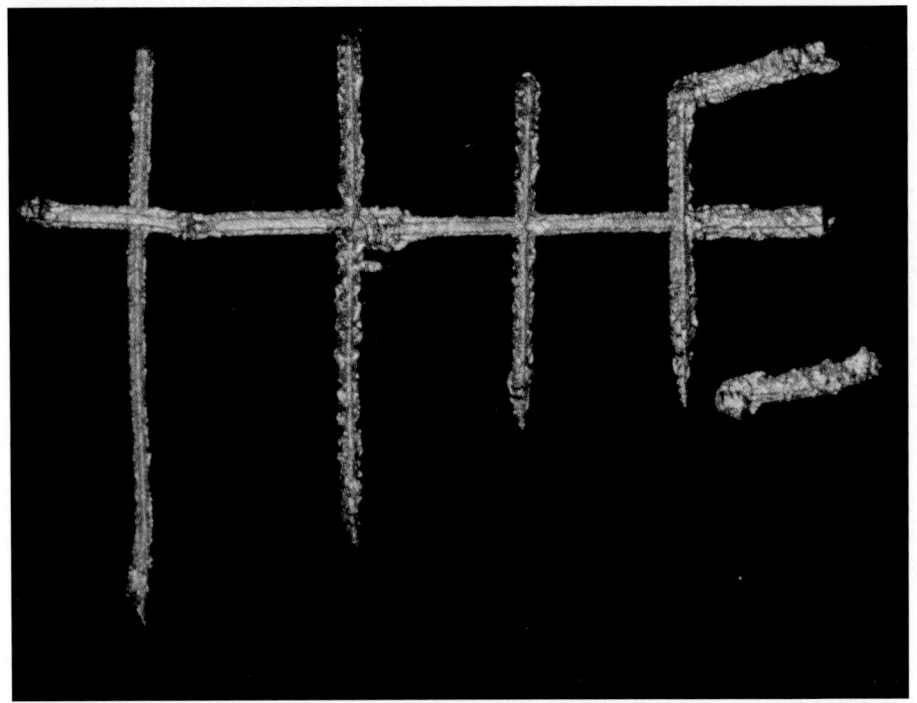

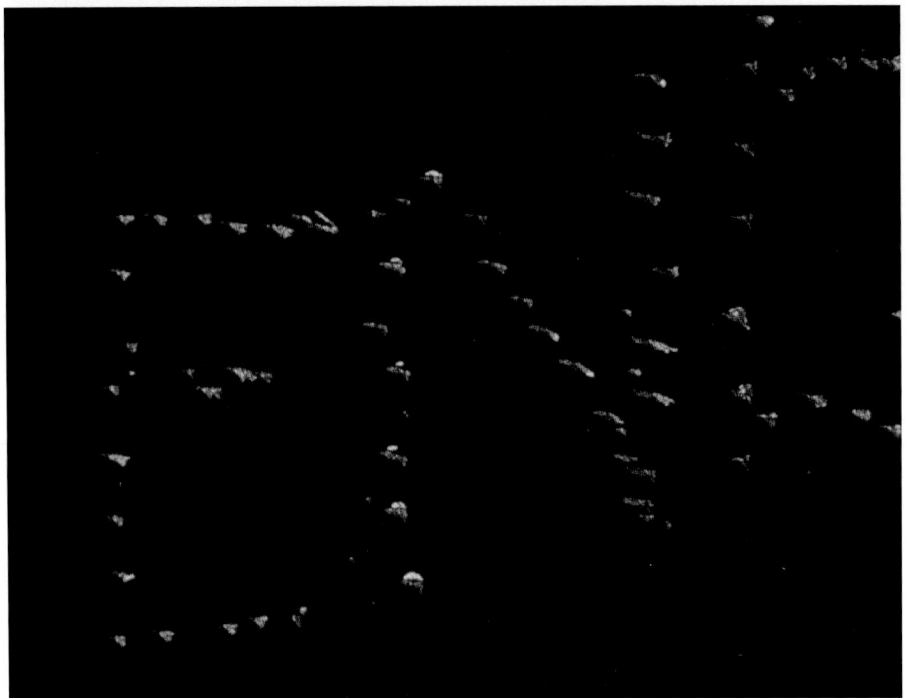

Len Lye, stills from *Particles in Space*, 1980.
Photographs courtesy Ngā Taonga Sound
& Vision.

STEVE RODEN

Los Angeles–based artist Steve Roden's works consider the relationship between sound and space to create a sense of wonder. Roden often translated vast and complex understandings of the world into quietly evocative objects, texts, sounds, and paintings. Known for his use of the term "lowercase" to describe his interest in celebrating the power of the hushed, the overlooked, and the incidental, Roden created a vast oeuvre that speaks to his curiosity and fascination with the world around him.

Sonic vibrations guided many of Roden's works; the artist often translated vibrational effects across media to create pieces that seemingly exist in a liminal state between movement and rest. *ear(th)* (2004) typifies this approach. The work uses seismic data as a score, prompting electronic arms to strike single xylophone bars that sit aloft an enormous custom-made hollow wood structure. Each chime heard is its own tiny vibrational event resonating on the surface of the world Roden invites viewers to step inside of. *ear(th)* was originally curated by Stephen Nowlin at ArtCenter College of Design in Pasadena, California, and developed in collaboration with scientists AnnMarie Thomas and Mark Simons. <

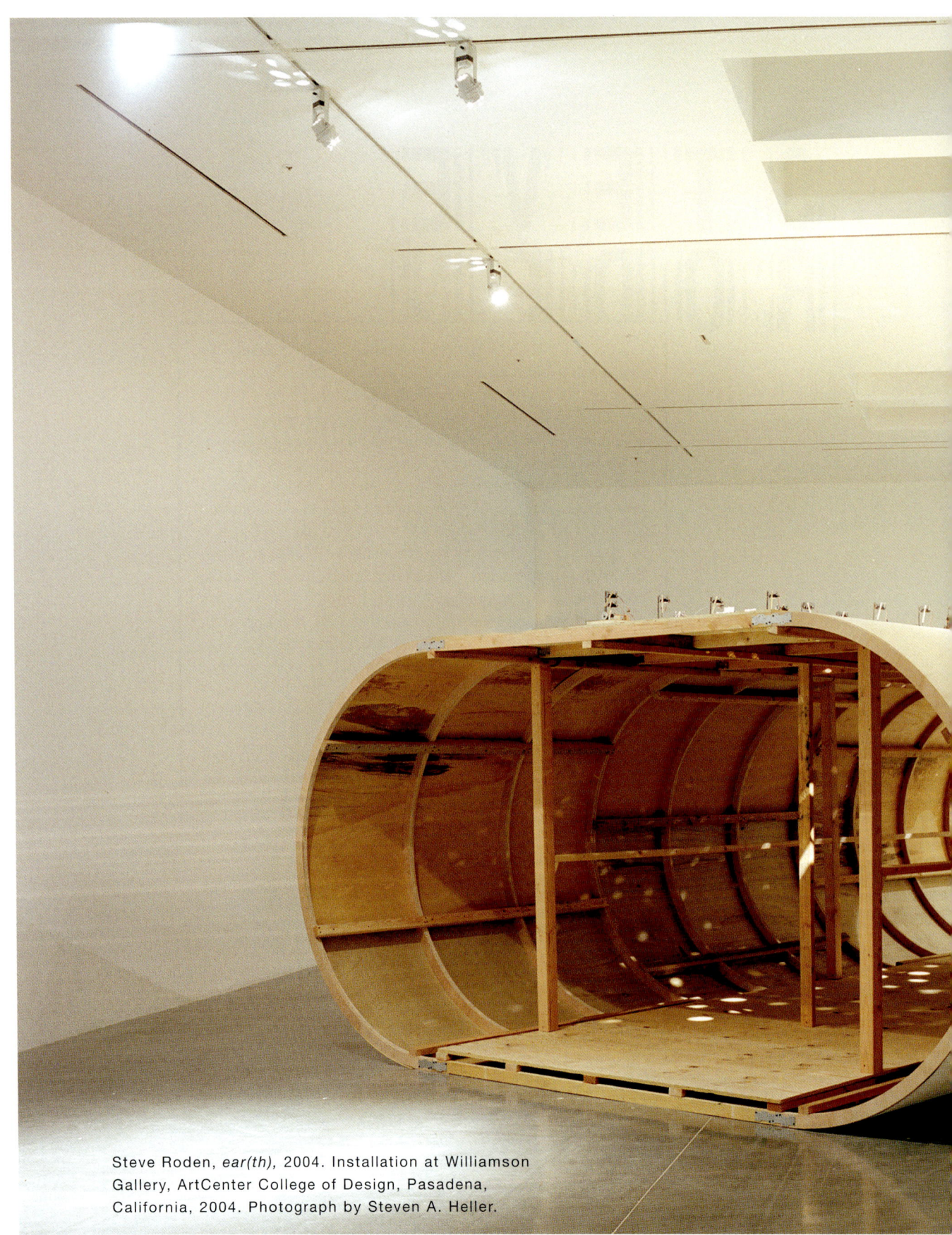

Steve Roden, *ear(th),* 2004. Installation at Williamson
Gallery, ArtCenter College of Design, Pasadena,
California, 2004. Photograph by Steven A. Heller.

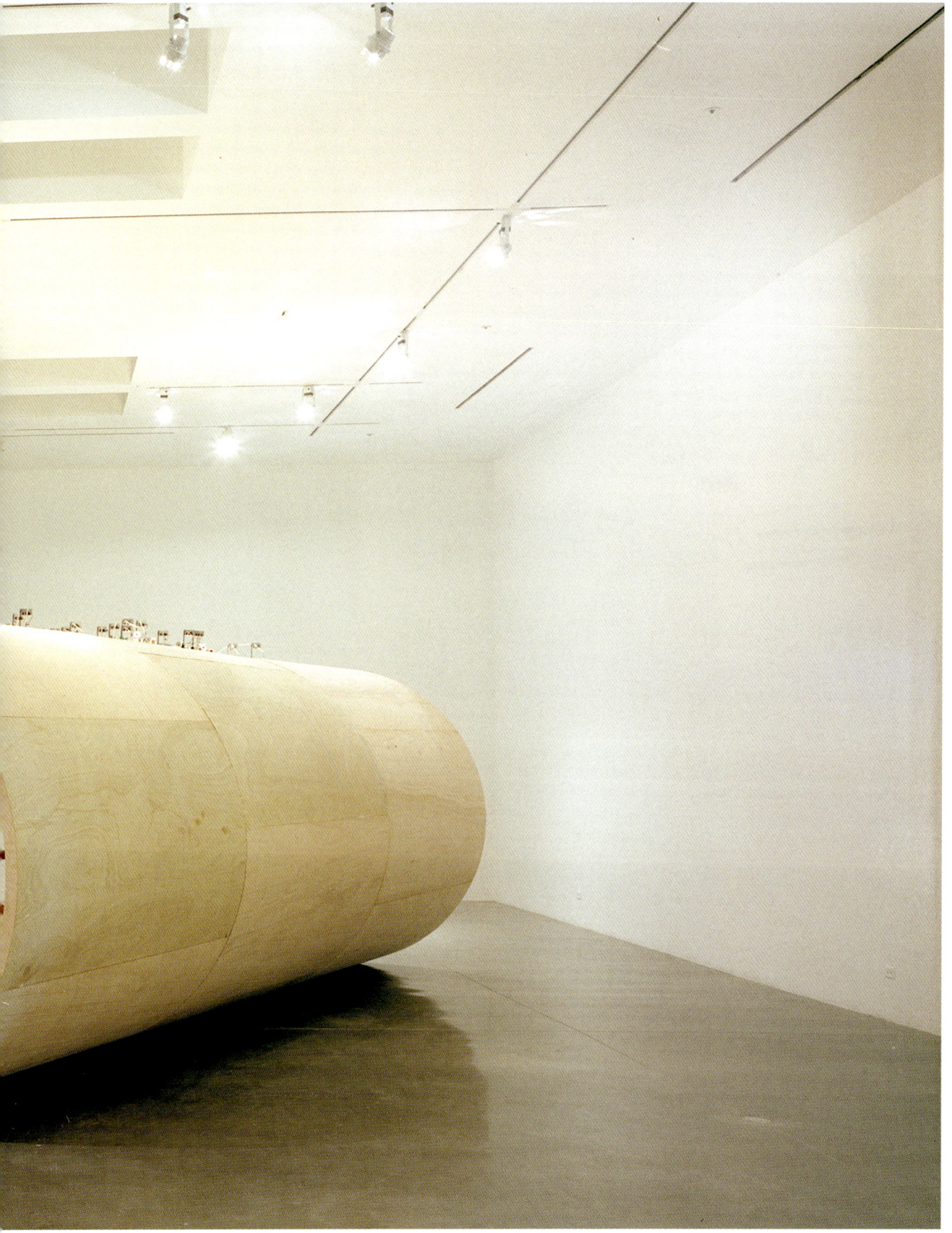

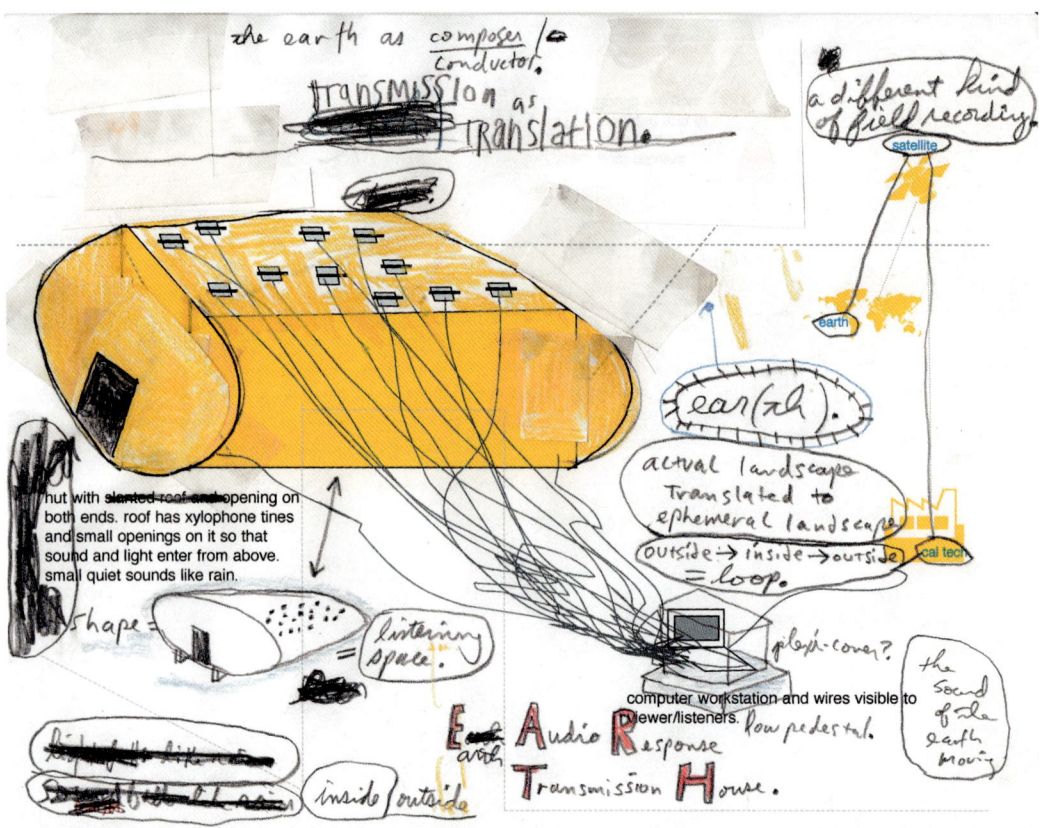

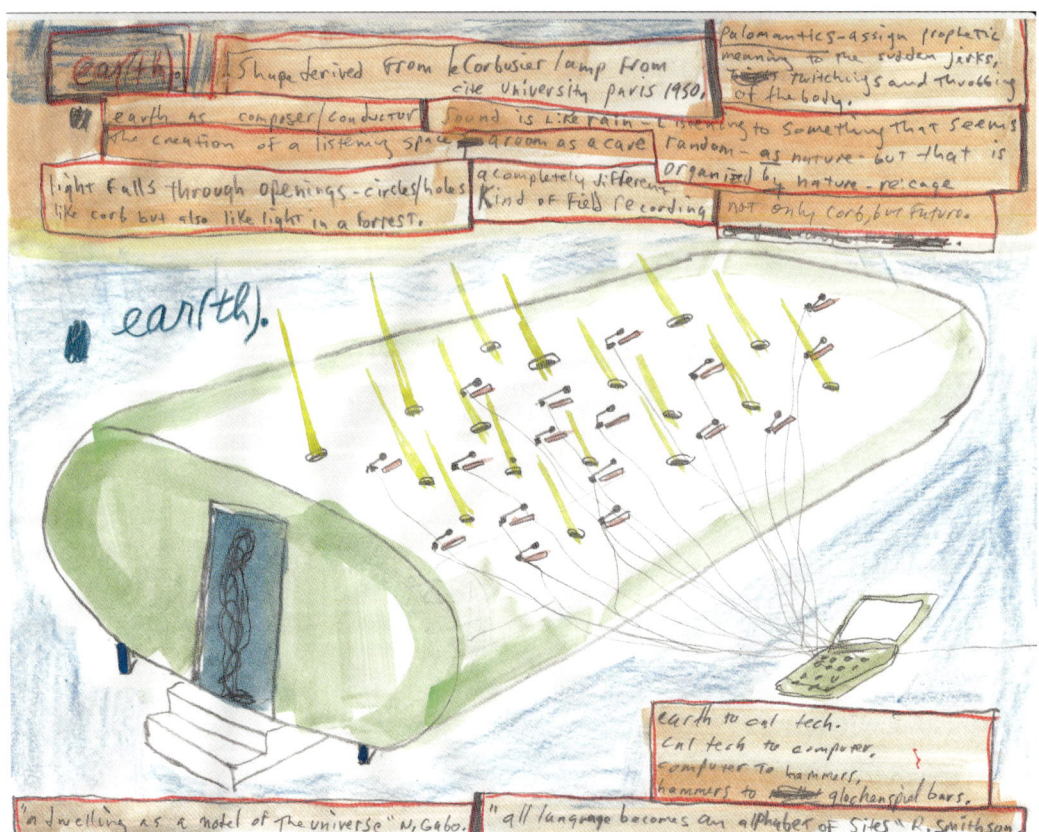

Steve Roden, concept drawings for *ear(th)*, c. 2004. Photographs © steve roden studio.

CHANNING HANSEN

Channing Hansen is a polymath who draws inspiration from the inherent beauty he finds in biological processes, genetic patterns, chaos theory, fractals, and the Fibonacci sequence. His artworks often take the form of vibrantly colored framed textiles he handknits according to algorithmic plans generated by a custom software program. These intricate plans echo the precision of DNA sequencing, and Hansen's translation of them into textiles mixes traditional artistic methods and craftsmanship with scientific principles foundational to our understanding of the world.

Energy Fields: Vibrations of the Pacific features a knitted textile work by Hansen, *Cosmic Fourier Fabric* (2023), that gives form to the artist's research on cosmic microwave background radiation (CMBR), understood as the afterglow of the big bang, or the residual heat of creation. To determine the work's composition, Hansen used NASA's 3-D modeling data of CMBR from the Wilkinson Microwave Anisotropy Probe, the Planck probe, and the Fourier transform equation to seed an algorithm. Knitted into a pattern that corresponds to the algorithm, *Cosmic Fourier Fabric* models an infinite-dimensional topology: space-time as a literal fabric, or the big bang as canvas. <

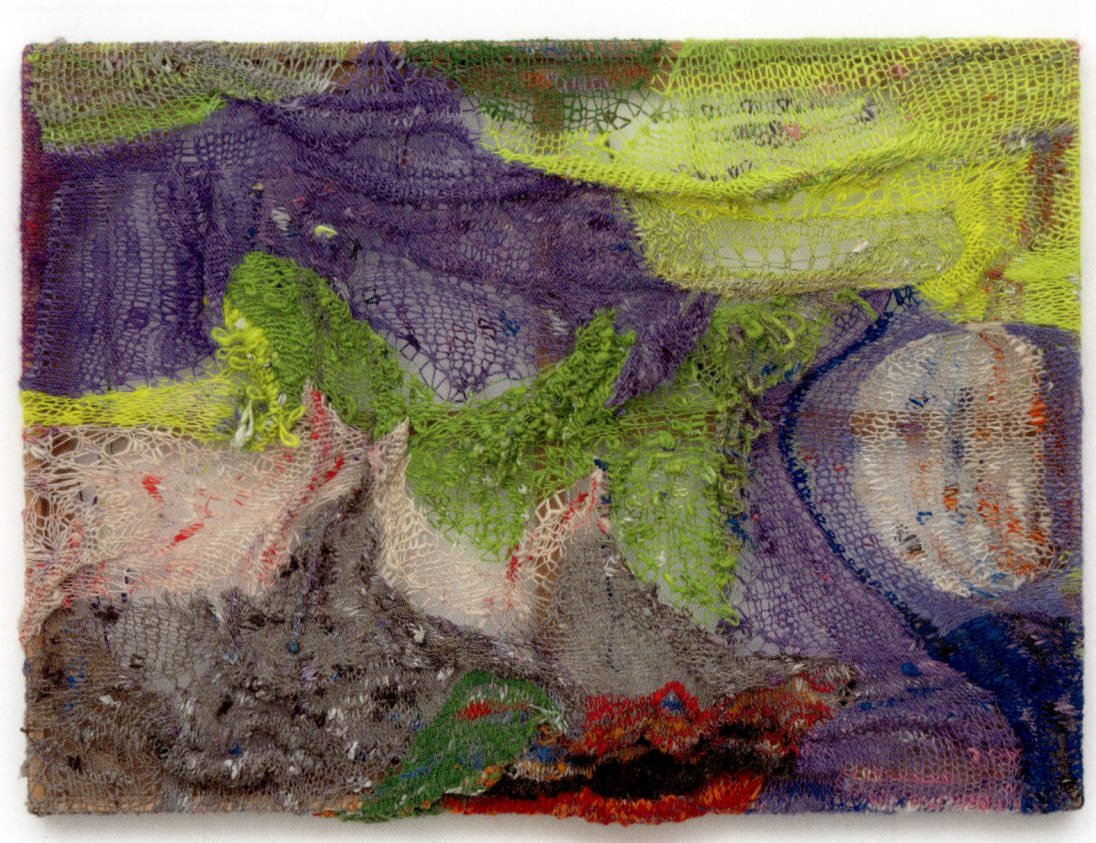

Channing Hansen, *7-Manifold*, 2017.
Alpaca, Bluefaced Leicester, California Variegated Mutant (Millie), California Variegated
Mutant (Petra), California Variegated Mutant
(Pine), cashmere, Churro, Exmoor Blueface, Lincoln, Lionhead (Beatrix and Derek),
Romeldale (January), Romeldale (Pallas),
Romney (O'Connor), Romney (Osiris), Romney (Martin), Romney (McKenna), Romney
(Noble), Romney (Oyster), Shetland (Shaun),
and Wensleydale fibers; casein, Bombyx
silk, silk noils, and Tussah silk; gold, holographic polymers, pearl dust, and photo-
luminescent recycled polyester fibers; and
banana cellulose, bamboo, bamboo carbon
fiber, legume cellulose, rose cellulose,
SeaCell, and Sequoioideae Redwood,
34 x 48 in.

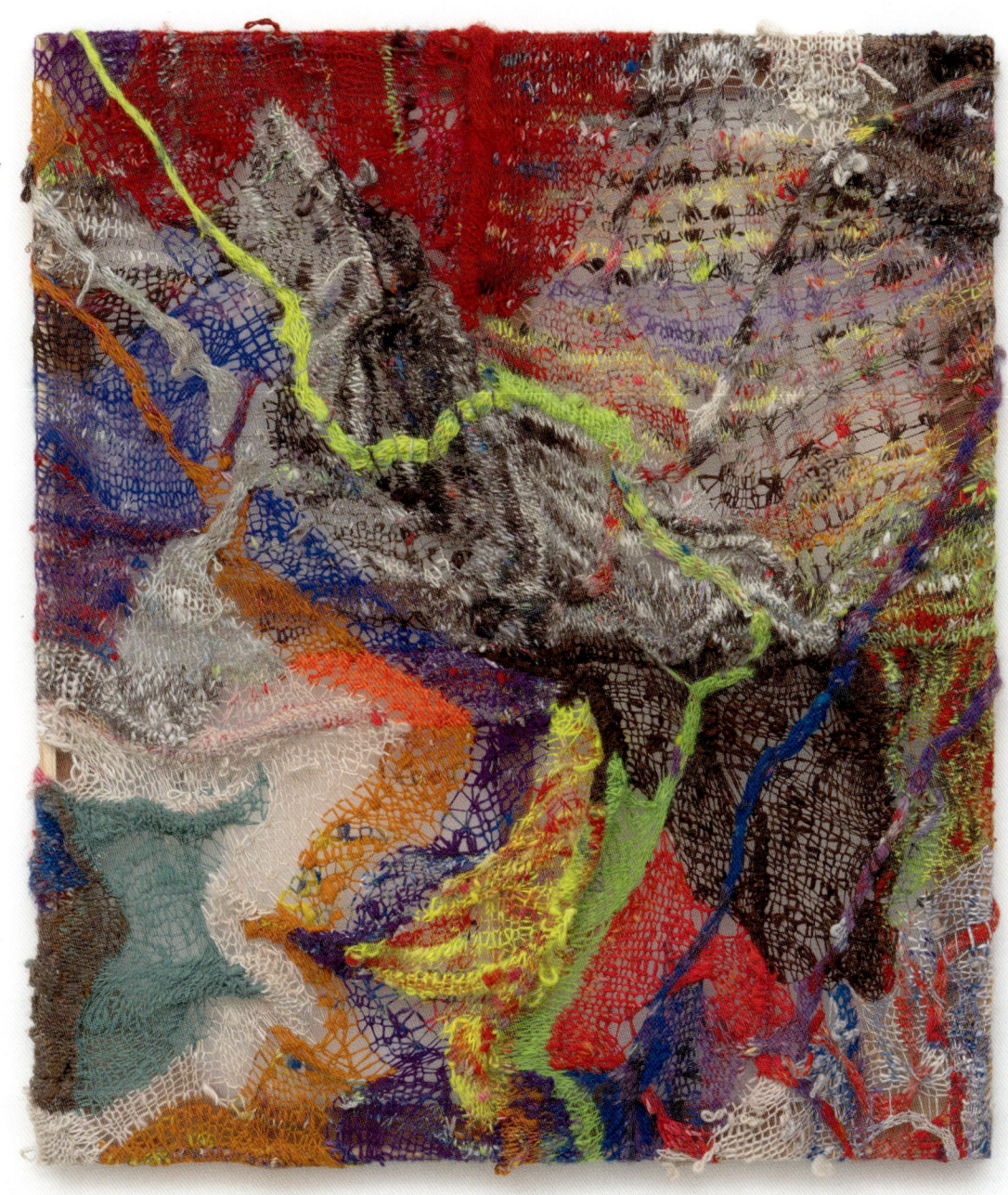

Channing Hansen, *9-Manifold*, 2017.
Bluefaced Leicester, California Variegated Mutant (Hattie), California
Variegated Mutant (Hope), California Variegated Mutant (Petra),
California Variegated Mutant (Pine), Dorset Horn, Exmoor Blueface,
Romeldale (January), Romeldale (Patty), Romney (Martin), Romney
(McKenna), Romney (Nevaeh), Romney (Noble), Romney (O'Connor),
Romney (Osiris), Romney (Princess), and Shetland (Freya) fibers;
silk noils and Tussah silk fibers; gold, holographic polymers, pearl
dust, and photoluminescent recycled polyester; and banana cellulose,
bamboo, bamboo carbon fiber, rose cellulose, SeaCell, legume
cellulose, and Sequoioideae Redwood, 42 x 48 in.

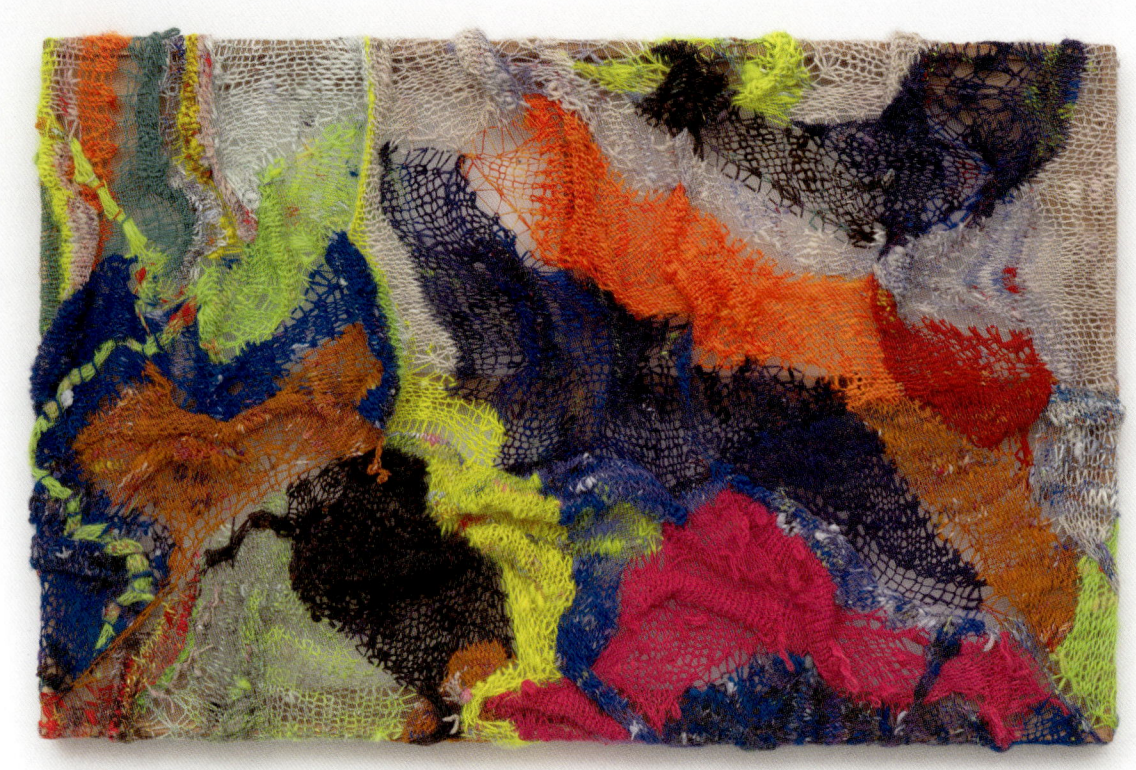

Channing Hansen, *10-Manifold*, 2017.
Bluefaced Leicester, California Variegated
Mutant (Hattie), California Variegated Mutant
(Hope), California Variegated Mutant
(Petra), California Variegated Mutant (Pierson),
California Variegated Mutant (Pine), Exmoor
Blueface, Romeldale (January), Romeldale
(Maggie), Romeldale (Palace), Romeldale
(Polly), Romney (McKenna), Romney (Martin),
Romney (Noble), Romney (Osiris), and Shet-
land (Freya) fibers; silk noils and Tussah silk
fibers; gold, holographic polymers, pearl
dust, and photoluminescent recycled poly-
ester; and banana cellulose, bamboo,
bamboo carbon fiber, rose cellulose, Sea-
Cell, legume cellulose, and Sequoioideae
Redwood, 32 x 51 in.

WILLIAM BASINSKI

William Basinski is a Los Angeles–based experimental sound artist renowned for his pioneering ambient music, soundscapes, and tape-loop compositions. Born in 1958, he has spent decades exploring the ephemeral nature of sound, memory, and decay. Basinski's most acclaimed project, *The Disintegration Loops* (2002–03), is a four-album series that captures the sounds of magnetic tape loops deteriorating as they are played again and again; Basinski conceived the work as an elegiac response to the events of September 11, 2001. For *Energy Fields: Vibrations of the Pacific*, Basinski was invited to perform his work *On Time Out of Time* (2019) in partnership with the Center for the Art of Performance at University of California, Los Angeles.

Basinski constructed *On Time Out of Time* from recordings in the Laser Interferometer Gravitational Wave Observatory (LIGO) archives that physicists have used to study the collision of two distant black holes over a billion years ago. Developed in conversation with artists Evelina Domintch and Dmitry Gelfand and experimental physicist Rana X. Adhikari, *On Time Out of Time* transforms scientific phenomena into a meditative auditory experience, inviting listeners to reflect on the vastness of the cosmos and the intricate fabric of time itself. <

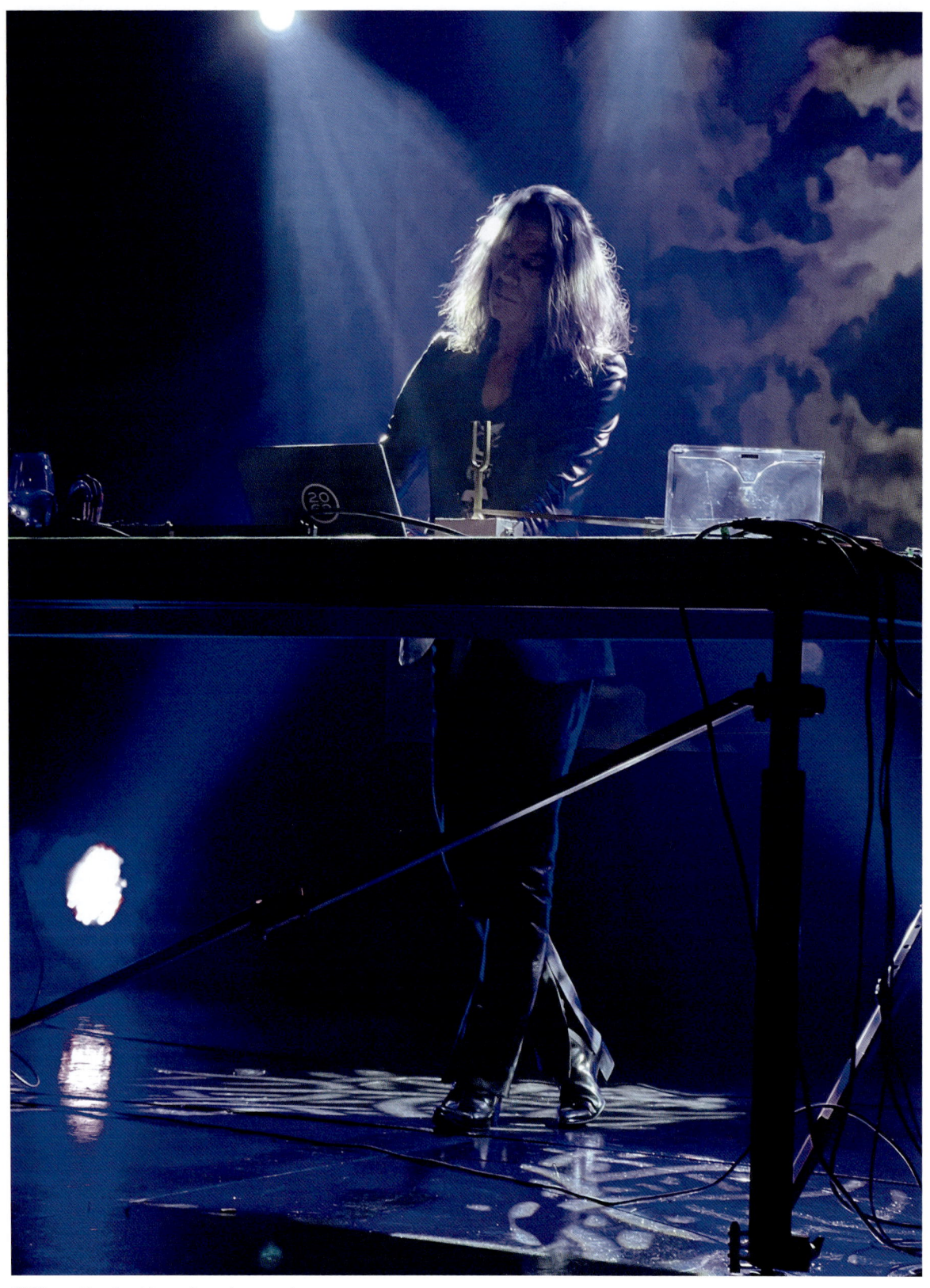

William Basinski performing, n.d.

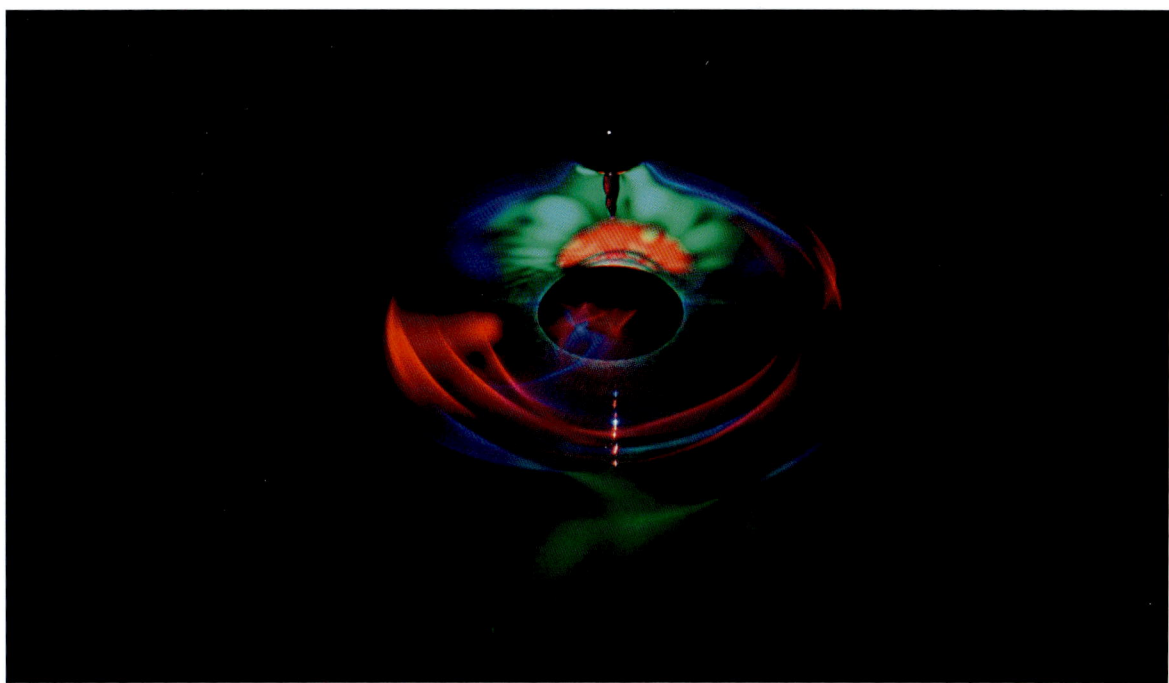

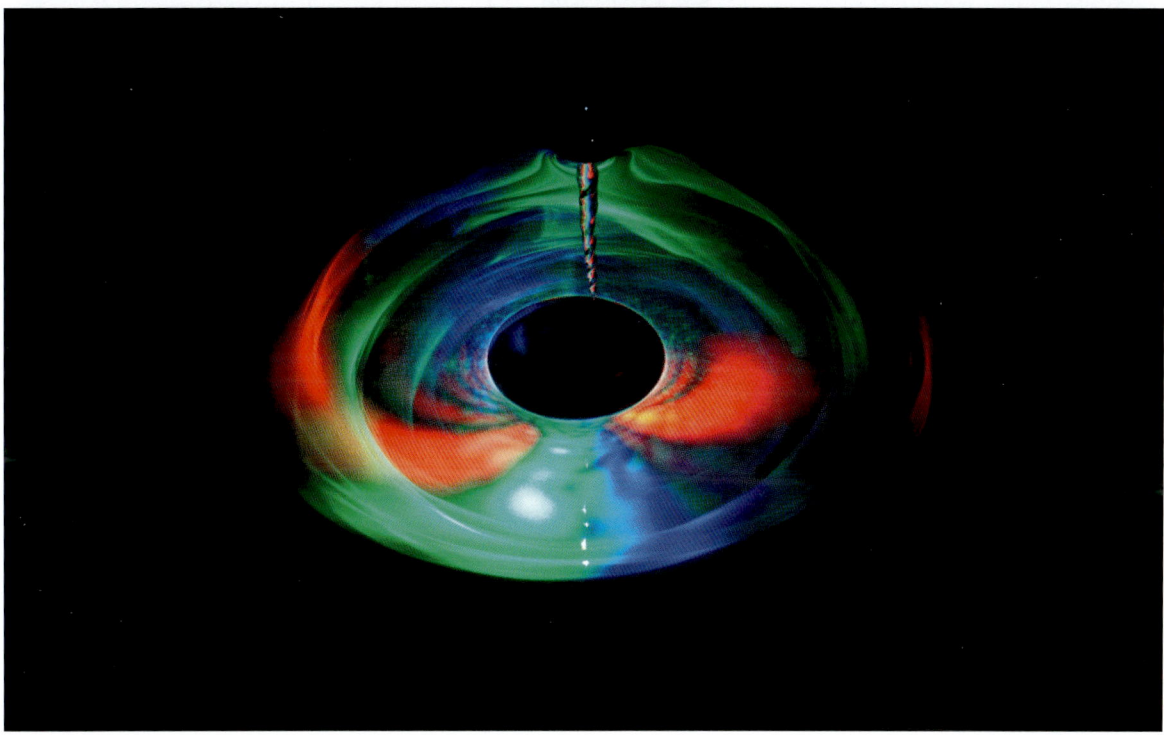

Evelina Domnitch and Dmitry Gelfand, *Orbihedron*,
2017, in collaboration with LIGO and William Basinski,

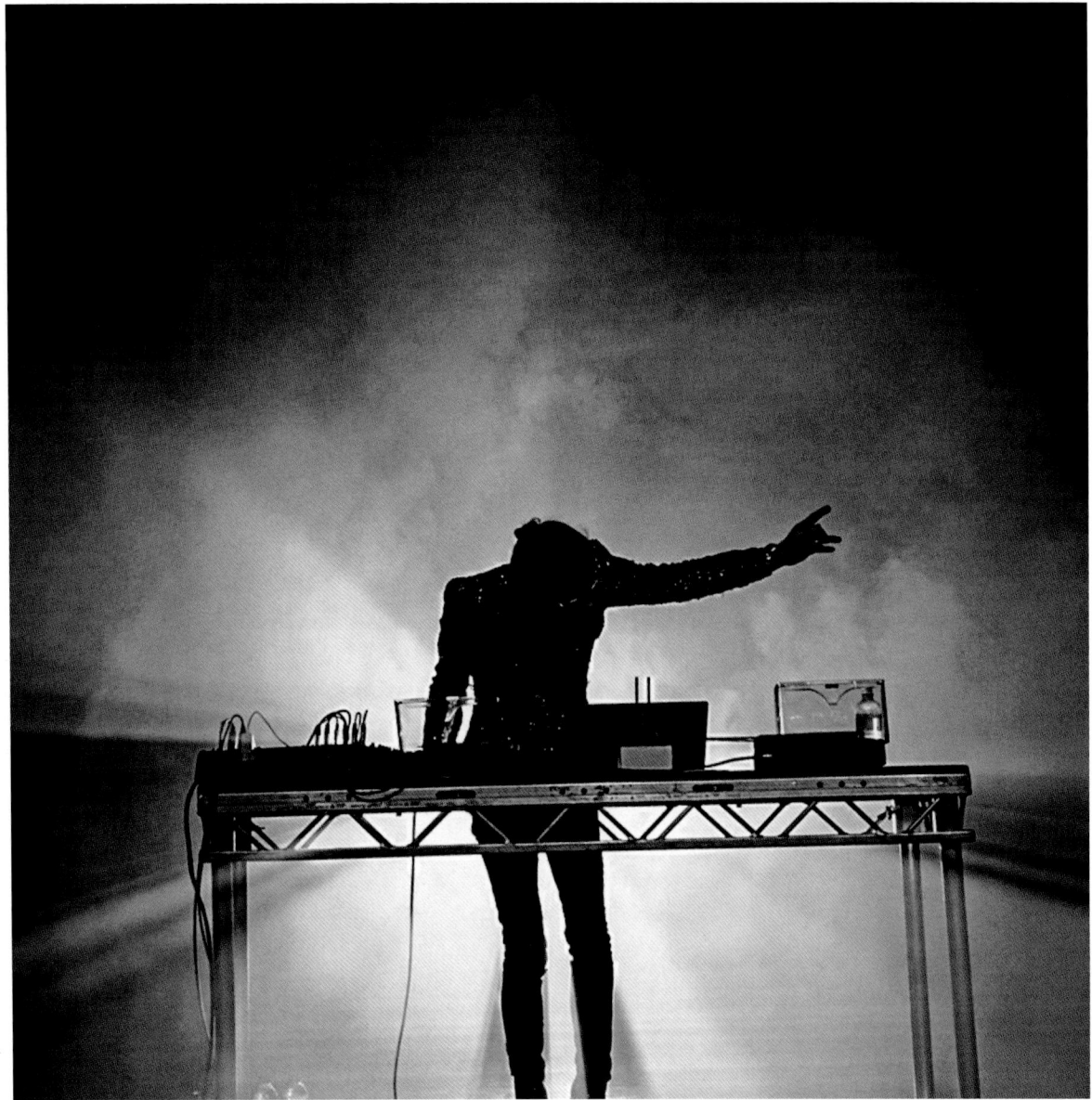

William Basinski performing, n.d.

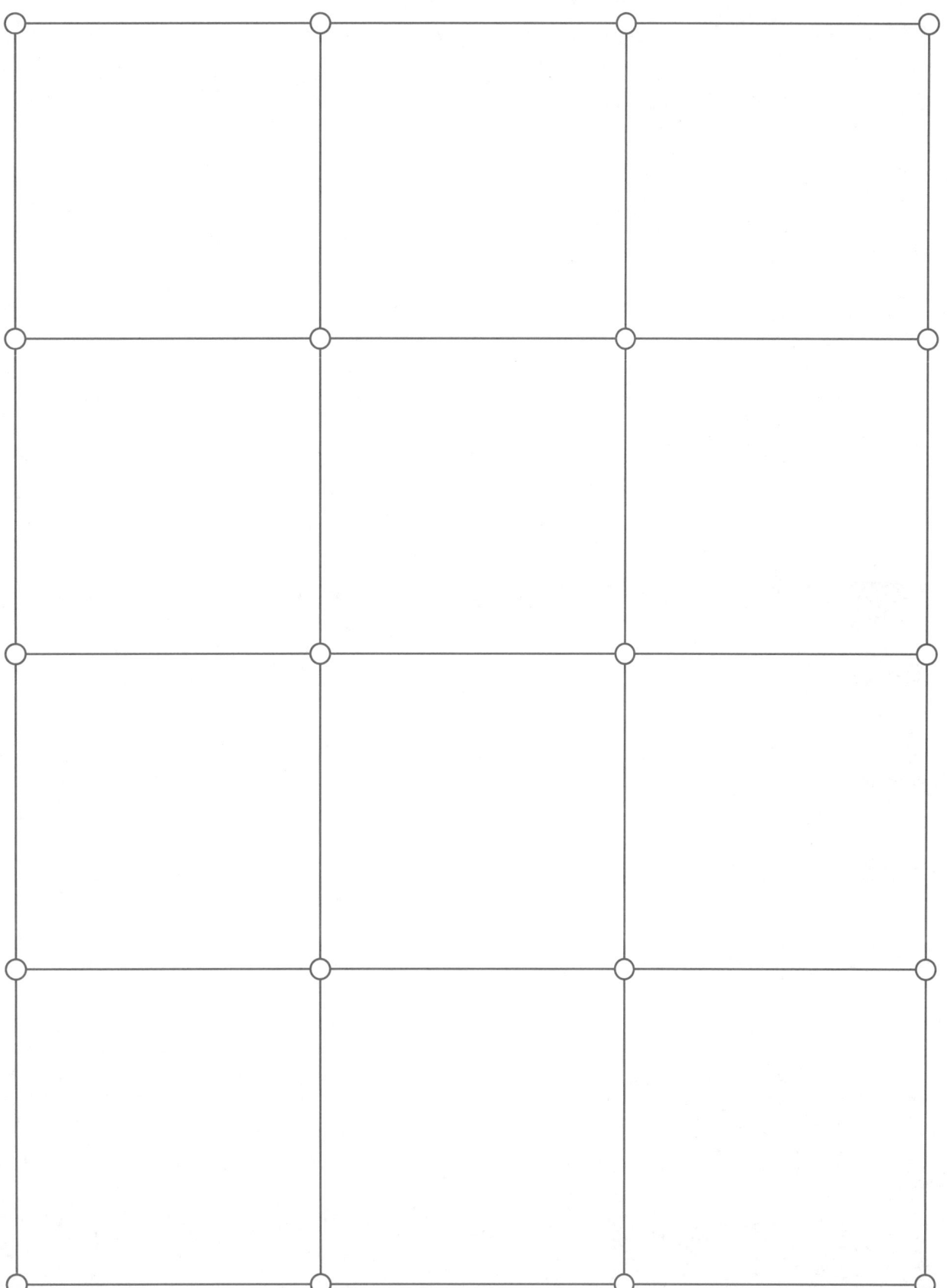

RANA X. ADHIKARI

Rana X. Adhikari is a professor of physics at the California Institute of Technology (Caltech) in Pasadena, California, and an associate faculty member of the International Centre for Theoretical Sciences in Bengaluru, India. He is among the scientists responsible for the Laser Interferometer Gravitational-Wave Observatory that discovered gravitational waves in 2015. At Caltech, he oversees the Adhikari Research Group, which investigates experimental gravitational physics related to gravitational wave astronomy and tests fundamental physics on the tabletop to understand the nature of spacetime.

LAWRENCE ENGLISH

Lawrence English is an artist, composer, and curator based in Australia. English uses a variety of approaches, including performance and site-specific installations, to invite audiences to consider their relationship to listening, place, and embodiment. Over the past decade, he has worked with vibration as a medium for sense making and is especially dedicated to the transformative and sensorial transgressive possibilities of sound. In 2015, he was awarded a Sidney Myer Creative Fellowship, and in 2022, he was honored by the Institute of Modern Art, Queensland. English earned his PhD from Queensland University of Technology in Australia.

MARCUS HERSE

Marcus Herse is the director of the Guggenheim Gallery at Chapman University in Orange, California. Originally from Germany, he is an artist, curator, and educator whose multidisciplinary investigations center on the experience of space and time as (dis)continuous. A graduate of the Kunstakademie Düsseldorf, Herse conducted his postgraduate studies at the University of California, Los Angeles. He has exhibited internationally, and his work has been reviewed in *Artforum*, *Flash Art*, and the *Los Angeles Times*.

VANESSA KWAN

Vanessa Kwan is director/curator of the Libby Leshgold Gallery at Emily Carr University of Art and Design in Vancouver. An artist, producer, and curator with a focus on collaborative, site-specific, and cross-disciplinary practices, they have held artistic leadership roles since 2003 and write, speak, and publish regularly on art and culture. Since 2017, they have been producing residency projects across the Pacific Rim exploring artist-led creative exchange.

W. PATRICK MCCRAY

W. Patrick McCray is a professor in the Department of History at the University of California, Santa Barbara (UCSB). Originally trained as a scientist, McCray's research focuses on the history of science and technology, and he has a courtesy appointment with UCSB's Media Arts and Technology Program. In 2020, McCray published *Making Art Work: How Cold War Engineers and Artists Forged a New Creative Culture* with MIT Press, which focuses on the activities and experiences of engineers and scientists who collaborate with artists.

ROBERT TAKAHASHI NOVAK

Robert Takahashi Novak (née Crouch) is the executive and artistic director of Fulcrum Arts in Pasadena, California. His curatorial work focuses on the overlapping disciplines of sound, science, technology, movement, and performance. Prior to joining Fulcrum Arts, Novak was associate director/curator at Los Angeles Contemporary Exhibitions. In 2017, he co-curated *Juan Downey: Radiant Nature* as part of the Getty Foundation's Pacific Standard Time: LA/LA. He is also a founding partner of VOLUME, a curatorial collective that functions as a catalyst for interdisciplinary new media work through exhibitions, performances, events, lectures, and publications.

PATRICK J. REED

Patrick J. Reed is a curator at Fulcrum Arts in Pasadena, California. Reed writes widely about contemporary art and culture with an interest in the language and visual representation of ecological disaster and the perversity of the sublime. At Fulcrum Arts, he oversees the online transmedia publishing platform, *Sequencing*, and conducts a research program exploring monumental works of "vanity engineering." His writing has been published in *ArtReview*, *ArtReview Asia*, *The Brooklyn Rail*, *Canvas: Art and Culture from the Middle East and Arab World*, *cura.*, and *e-flux Criticism*.

ENRIQUE RIVERA

Enrique Rivera is the executive director of the Museo Interactivo Mirador in Santiago, where he explores the relationship between art, science, and the environment through exhibitions and special projects. From 2013 until 2022, Rivera was president of the Chilean Corporation of Video and Electronic Arts, where he oversaw the Media Arts Biennial and the Juan Downey International Media and Audiovisual Arts Contest. Rivera studied cinema at the Universidad de Chile and art direction at the Universidad Mayor in Santiago.

RACHEL SHEARER

Rachel Shearer (Rongowhakaata, Te Aitanga a Māhaki, Pākehā), based in Aotearoa (New Zealand), is interested in listening to the energies of the earth through Western and Māori philosophies and technologies. Through her practice, she variably engages experimental music, field recordings, embodied listening, installation, sound/spatial design, moving image, and writing towards this *kaupapa* (theme).

FIONA SHEN

Fiona Shen is the Escalette Collection Director at Chapman University in Orange, California. Prior to this, Shen was associate professor and director of research at Sino-British College in Shanghai. She is the author of *Pearl: Nature's Perfect Gem* (Reaktion Books, 2022), *Silver* (Reaktion Books, 2017), and *Knowledge is Pleasure: Florence Ayscough in Shanghai* (Hong Kong University Press, 2012), and her writing has appeared in *Orion*, *The Washington Post*, *China Today*, and *The Cimarron Review*. Shen received a PhD in art history from the University of St. Andrews, Scotland.

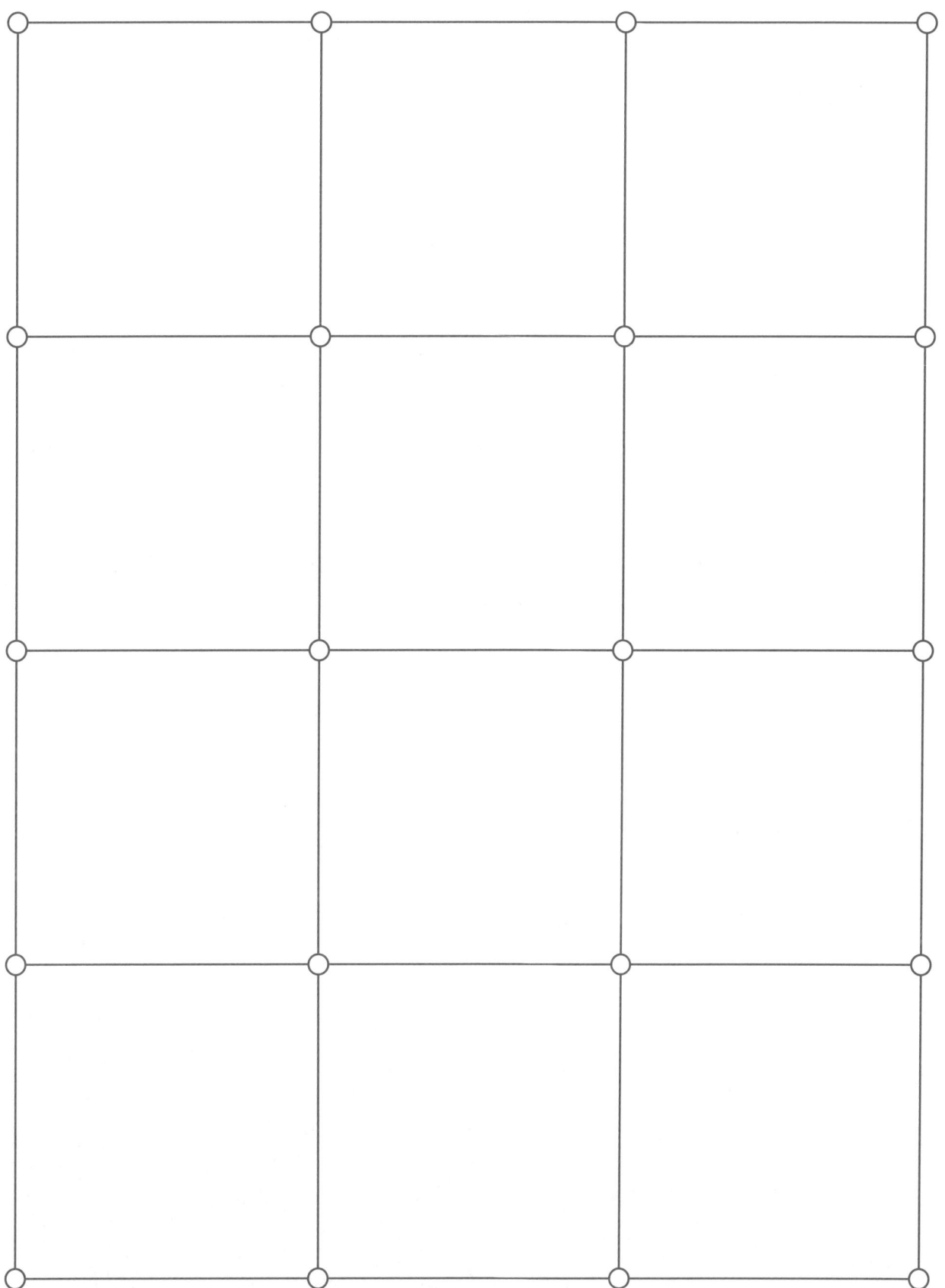

ACKNOWLEDGMENTS

Energy Fields: Vibrations of the Pacific is dedicated to the memory of Steve Roden.

The creation and realization of this exhibition would not have been possible without the generous support of the Getty Foundation. We extend our deepest gratitude to Joan Weinstein, director, and Heather MacDonald, senior program officer, for their unwavering commitment; their guidance and encouragement were instrumental in bringing this exhibition to life. We also extend our heartfelt thanks to Getty Foundation staff Selene Preciado, Zachary Kaplan, and Lu Spriggs for their essential support and contributions.

We extend heartfelt thanks to our collaborators at Chapman University, especially Marcus Herse, director of the university's Guggenheim Gallery, for his friendship, camaraderie, and support. We are also deeply grateful to Fiona Shen, director of the university's Escalette Permanent Art Collection, for facilitating the exhibition's educational programming, and to Jessica Bocinski, collections manager, for deftly handling the registration of artworks. Special thanks go to Lia Halloran, associate professor of art and chair of Chapman's Department of Art, for introducing Fulcrum Arts to the university, facilitating this fruitful partnership.

Energy Fields: Vibrations of the Pacific benefited tremendously from the incredible team at Fulcrum Arts, who played a critical role in every phase of development. First and foremost, this exhibition would not have been possible without the unwavering support and commitment of the Fulcrum Arts Board of Trustees, especially Board President Scott Tennent. In addition, Fulcrum Arts Curator and Artist Program Manager Patrick J. Reed had the monumental task of managing both the publication and exhibition, Digital Media and Marketing Assistant Scarlett Wang created the bold marketing vision, and Grants Writer Karen Lofgren assisted with research and helped secure early funding. We are grateful to our excellent publication team, editor Elizabeth Hamilton, designer Willem Henri Lucas, and Set Margins' publisher Freek Lomme, who collaborated with us to produce this beautiful catalogue. We are also grateful to Marie Bland, Martin Carrillo, Nick Cimiluca, Emma Jacobson-Sive, Jana Juhl, Ed Patuto, Geneva Skeen, Sarah Stifler, Jaspa Ureña, Tyler Wert, Holly Witham, and Christopher Wormald for their invaluable contributions.

The development of *Energy Fields: Vibrations of the Pacific* was supported by a stellar group of researchers, advisors, and program partners, including Rana X. Adhikari, Marcus Herse, Vanessa Kwan, W. Patrick McCray, Enrique Rivera, Fiona Shen, and Nina Tonga. In addition, Sam Rowell conducted deep research into the exhibition's overlapping themes and created a stunning sound series for Lookout FM. Joel Ferree, Annea Lockwood, Perrin Meyer, Kyle Slabb, and Alex Wellerstein lent critical guidance and support, and Jay Marx and Adriel Luis helped with early research. During our 2023 research trip to Japan, we were generously hosted by Haco, Aki Onda, and Akio Suzuki, and we are particularly grateful to Hiromi Miyakita and Minoru Sato for making our experiences of Tottori and Kamakura truly special and unforgettable. We also acknowledge Megan Denz and Paul Brobbel of the Len Lye Centre and Steve Russell of Ngā Taonga Sound & Vision for their guidance and encouragement.

We are also deeply grateful to our programming partners, who have enabled us to fully realize *Energy Fields: Vibrations of the Pacific*: Johanna Burton, director of the Museum of Contemporary Art, Los Angeles, and her staff of Amelia Charter, Michele Huizar, Clara Kim, and Alex Sloane; Edgar Miramontes, executive and artistic director of the Center for the Art of Performance at the University of California, Los Angeles, and his team of Bozkurt "Bozzy" Karasu, Katelan Braymer, Fred Frumberg, and Duncan Woodbury; Timothy Phillips, superintendent of the Los Angeles County Arboretum Botanical Garden, and Brooke Applegate, Nadia Balaz, Larry Giannone, Rachel Masters, Sylvia Rosenberger, Jennifer Van Stralen, and Angela Zhu; and Tim Leanse and his team at Zebulon.

The Getty's PST ART: Art & Science Collide initiative gave us the unique opportunity to present *ear(th)* (2004), a major artwork by the late artist Steve Roden. We are deeply grateful to Sari Roden for trusting us to realize his vision, and to Meg Linton for her ongoing support, research, and guidance. We are also deeply grateful to Jeff Kleeman and his team at Studio Sereno for faithfully re-creating this ambitious artwork, and to AnnMarie Thomas for consulting with us about her scientific and engineering contributions to the original presentation. In addition, we thank Rebecca McGrew, Stephen Nowlin, John David O'Brien, Susanne Vielmetter, and Stephen Vitiello for their friendship, guidance, and support. We also recognize Mary Clare Stevens, executive director, and Rochele Gomez, grants manager, at the Mike Kelley Foundation for providing critical support for the project.

We are grateful to the National Endowment for the Arts for their generous support of the performance program and Fulcrum Festival, as well as to Kristin Sakoda, director,

arts and culture, and Anji Gaspar-Milanovic, director, grants and professional development, at the Los Angeles Department of Arts and Culture, and to Christina Chu and her team at SOLARPUNKS for their in-kind support towards realizing Annea Lockwood's *Wild Energy* (2014) at the LA Arboretum. We also give special thanks to Zohar Spatz at Creative Australia. Last but not least, we extend our deepest gratitude to Stephania Ramirez and the Perenchio Foundation for providing the critical infrastructural support without which this project would not have been possible.

We are grateful to our families and friends for their encouragement and understanding throughout this endeavor. Lawrence extends special thanks to Rebecca, Frances, Theodore, and Augustine for their energy and support over the past years. Robert thanks his husband, Yann Novak, for his unending patience, generosity, and humor throughout the development of *Energy Fields*.

Lawrence English
Co-Curator, *Energy Fields: Vibrations of the Pacific*

Robert Takahashi Novak
Co-Curator, *Energy Fields: Vibrations of the Pacific*
Executive and Artistic Director, Fulcrum Arts

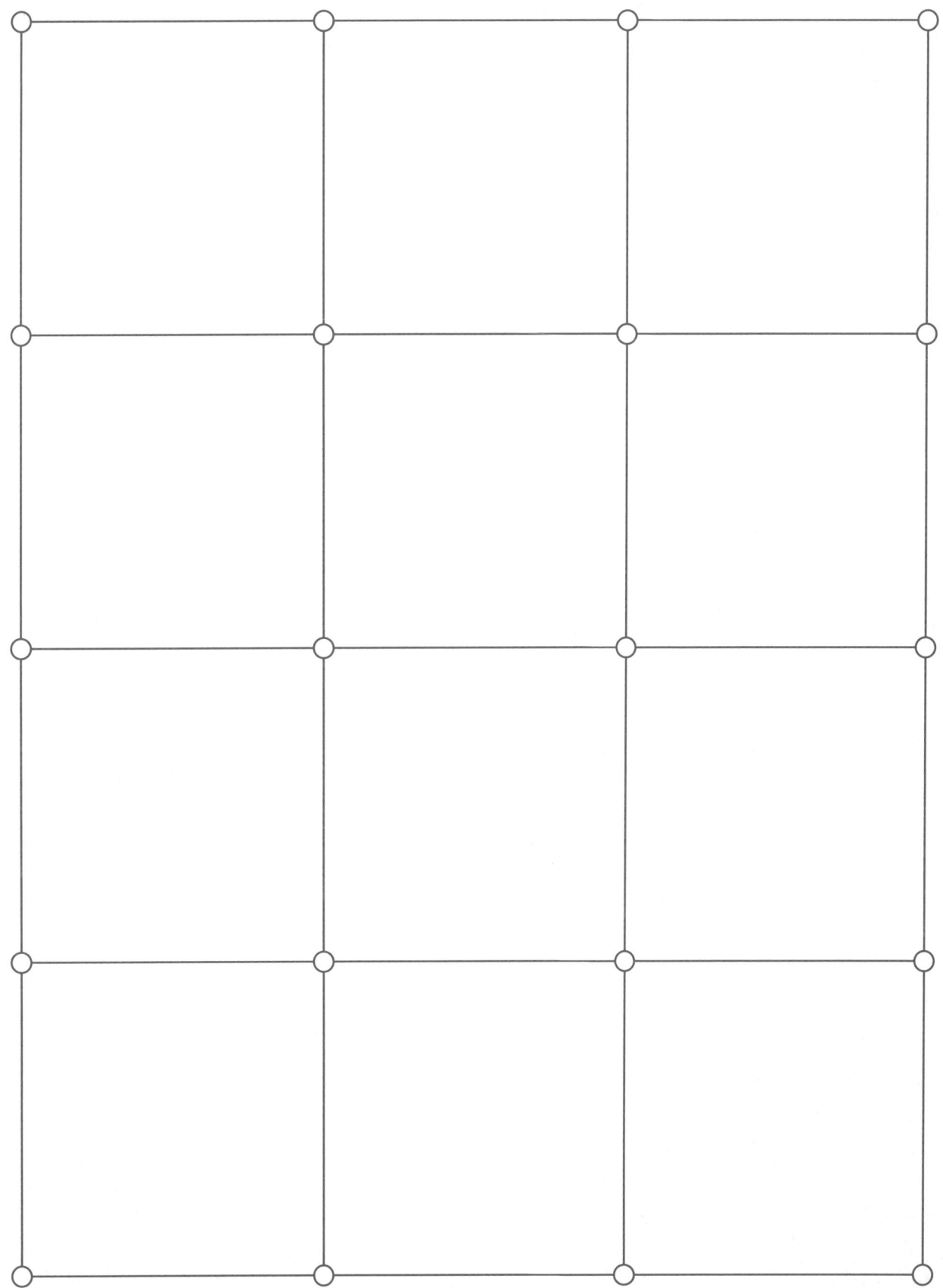

CHECKLIST OF THE EXHIBITION

**LAUREN BON AND
THE METABOLIC STUDIO**
(b. 1962, New Haven, Connecticut;
lives and works in Los Angeles)
The Great Vibration, 2023
Digital video; color and sound
13:00 minutes
Courtesy of the artist

JENEEN FREI NJOOTLI
(b. 1988, Whitehorse, Canada;
lives and works in Old Crow, Yukon
Territory)
*Bone meets blade, sonified calling
out. who what will utter back*, 2023
Caribou shoulder bone, Dremel,
microphone, audio power amplifier
with video, and audio documentation
of live performance
Dimensions variable
Courtesy of the artist and Macaulay
& Co. Fine Art

DAVID HAINES
(b. 1966, London)
and JOYCE HINTERDING
(b. 1958, Melbourne)
(Live and work in New South Wales,
Australia)
Telepathy, 2008–ongoing
Timber, anechoic tiles, acoustic
barrier rubber, plasterboard, video
monitors, and video transmitters
80 x 80 x 236 in.
Courtesy of the artists

CHANNING HANSEN
(b. 1972, Los Angeles; lives and
works in Los Angeles)
Cosmic Fourier Fabric, 2023
Californian Variegated Mutant
(Pierson), Cotswold (Sassy),
Romeldale (January), Romeldale
(Qassiopeia), Romeldale (Shelby),
and Romeldale (Tapestry) fibers;
silk noil and tussah silk; holographic
polymers and photo-luminescent
recycled polyester; bamboo carbon
fiber; Ingeo corn and pineapple fibers;
pearl-infused cellulose; soy silk; and
Sequoioideae Redwood

40 x 42 in.
Courtesy of the artist and
Marc Selwyn Fine Arts

*Mapping The Universe Popular Atlas:
2024 Edition*, 2024
Spiral bound and printed on 130 lb TC
Digital Satin (cover), 80 lb Mohawk
EDD Opaque Smooth Text (interior
maps), and 100 lb TC Digital Gloss
Text (interior artwork)
9 $^3/_4$ x 8 $^1/_4$ in.
Courtesy of the artist

VIRGINIA KATZ
(b. 1956, New York; lives and works
in Irvine, California)
*WIND, On-Shore Flow, 7 Hours
of Observation, Green and Blue,
3/28/08*, 2008
Metallic ink on black paper
44 $^1/_2$ x 30 in.
Courtesy of the Escalette Permanent
Art Collection at Chapman University

*WIND, Off-Shore Flow, 10 Hours
of Observation, Gold and Copper,
10/07/08*, 2008
Metallic ink on black paper
44 $^1/_2$ x 30 in.
Courtesy of the Escalette Permanent
Art Collection at Chapman University

ANNEA LOCKWOOD
(b. 1939, Aotearoa (New Zealand])
Wild Energy, 2014
Composition in collaboration
with Bob Bieleck;
46:00 minutes
Courtesy of the artist

LEN LYE
(b. 1901, Christchurch, New Zealand;
d. 1980, Warwick, New York)
Particles in Space, 1980
16mm film; black-and-white and sound
3:50 minutes
Digital version by Park Road Post
Production and Weta Digital Ltd from
material preserved and made available
by Ngā Taonga Sound & Vision
Courtesy of the Len Lye Foundation
Stills collection: Ngā Taonga Sound
& Vision

ROSS MANNING
(b. 1978, Brisbane, Australia; lives
and works in Brisbane)
Ambient Painting, 2016
Dichroic glass filters and natural and
artificial light
Dimensions variable
Courtesy of the artist and Milani
Gallery

STEVE RODEN
(b. 1964, Los Angeles; d. 2023,
Los Angeles)
ear(th), 2004
Wood, glockenspiels, robotic
elements, electric circuitry, and
looping audio
192 x 96 x 288 in.
Courtesy of the Estate of Steve Roden

MINORU SATO
(b. 1963, Sendai, Japan; lives and
works in Kamakura City, Japan)
Thermal Acoustics, 2013–ongoing
Glass tubes, Liebig condensers,
nichrome heaters, handmade
amplifiers, handmade timer,
loudspeakers, and microphone
96 x 96 x 120 in.
Courtesy of the artist

RACHEL SHEARER
(b. 1966, Aotearoa [New Zealand];
lives and works in Aotearoa)
Whakapapa of Shimmers, 2024
Super 8mm film; color and sound
8:00 minutes
Courtesy of the artist

KYLE SLABB
(b. 1974; lives and works in
Bundjalung country, Australia)
Binanggu, 2024
Ink on paper
39 $^3/_8$ x 20 $^1/_{16}$ in.
Courtesy of the artist

AKIO SUZUKI
(b. 1941, Pyongyang, North Korea;
lives and works in Kyōtango, Japan)
ANALAPOS–a, 1970
Metal tubes, custom designed spring,
and speaker
Dimension variable

Courtesy of the artist.

ANALAPOS–a, 2023
Video documentation of performance
at Tottori Sand Dunes, Japan;
color and sound
4:16 minutes
Courtesy of the artist

MALENA SZLAM
(b. 1979, Santiago; lives and works
in Montreal)
Altiplano, 2018
16mm transferred to 35mm film; color
and sound
15:30 minutes
Courtesy of the artist

ALBA TRIANA
(b. 1969, Bogotá; lives and works
in Miami)
Music on a Bound String No. 2, 2015
Visible sound and light, mechanical
and electronic elements
Dimensions variable
Courtesy Liza and Dr. Arturo F.
Mosquera

MO H. ZAREEI
(b. 1985, Tehran; lives and works
in Wellington, New Zealand)
Material Music, 2019–20
Wood, metal, glass, marble,
mechanical and electronic elements
36 x 12 x 5 in.
Courtesy of the artist

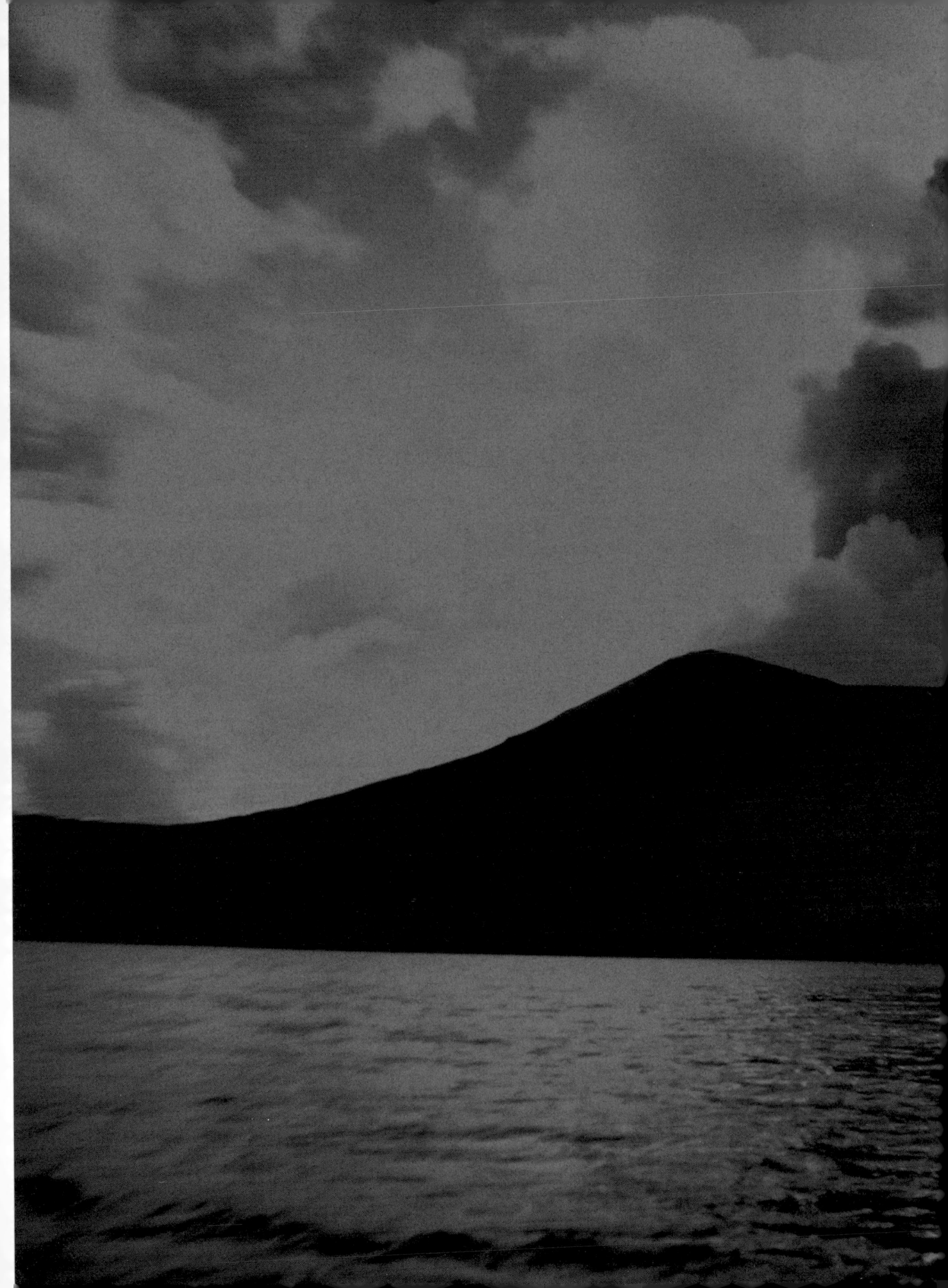